The Riemann zeta function and related themes

Ramanujan Mathematical Society Lectures Notes Series

Vol. 1: Number theory

Vol. 2: The Riemann zeta function and related themes

Vol. 3: Formal language aspects of natural computing

Vol. 4: Commutative algebra and combinatorics

Vol. 5: Convexity in discrete structures

Vol. 6: Number theory and discrete geometry

Vol. 7: Discrete mathematics

Vol. 8: Lectures on operator theory

Vol. 9: Essays on geometric group theory

Vol. 10: Techmüller theory and moduli problems

Ramanujan Mathematical Society

Lecture Notes Series

Volume 2

The Riemann zeta function and related themes

Proceedings of the international conference held at the National Institute of Advanced Studies, Bangalore, December 2003

Volume editors

R. Balasubramanian
K. Srinivas

International Press
www.intlpress.com

Ramanujan Mathematical Society
Lecture Notes Series, Volume 2
The Riemann zeta function and related themes

Volume Editors:
R. Balasubramanian
K. Srinivas

Published in 2010 by International Press, Somerville, Massachusetts, U.S.A. under license from the Ramanujan Mathematical Society.

This work is also published in the Republic of India, exclusively by the Ramanujan Mathematical Society.

ISBN 978-1-57146-187-2

Typeset using the LaTeX system.
Printed in the United States of America.

Foreword

As I have said on other occasions, in my opinion, K. Ramachandra is the real successor of Ramanujan in contemporary Indian mathematics. Following the tradition of Chowla, Pillai and Vijayaraghavan, he and his students have continued to make significant contributions to classical number theory. It is not surprising that active mathematicians from various countries came to attend the conference held at Bangalore in December 2003, to offer their felicitations to him on completing seventy years of meaningful life. I did not attend the conference, because I have been out of active mathematics for some time. It is, therefore, all the more gracious of the editors, R. Balasubramanian and K. Srinivas to give me the honour of writing this foreword to the conference proceedings. I not only admire Ramachandra for his mathematics, but also greatly value his friendship, and the regard and affection he has shown to me over the years.

Ramachandra, born in 1933, completed his B.Sc (Hons.) and M.Sc. studies at Bangalore. He got his Ph.D. in 1965 at TIFR under the supervision of K.G. Ramanathan, whose influence on Indian mathematics has been remarkable.

Ramachandra's initial work was on algebraic number theory. The reviewer M. Eichler, of his first paper: Some applications of Kronecker's limit formulas, Ann. of Math (2), 80 (1964), 104-148, started the review with the remark: "This paper contains some remarkable new results on the construction of the ray class field of an imaginary quadratic number field."

When the seminal work of Alan Baker appeared in the 1960's, he and his students, especially T.N. Shorey, took up transcendental number theory and made remarkable contributions to the Baker theory and its applications to problems of classical number theory.

Another important result proved by Ramachandra states: for m large, between m^2 and $(m + 1)^2$ there is an n and a prime p dividing n such the $p > n^{1/2+1/11}$.

After 1974, he turned his attention to hard core classical analytical number theory, especially Reimann zeta function and general Dirichlet series. His school has made significant contributions, especially to gaps between ordinates of critical zeros of the zeta function and mean values of the fractional powers of the Riemann zeta function inside the critical strip. The method he developed to give a simple proof for the asymptotic formula for the fourth power mean and his application of a contour suggested by Huxley and Hooley has found wide applications.

With his contributions, it was natural for him to receive various honours like election to fellowships of various science academics, medals and award lectures, especially Ramanujan medals of Indian National Science Academy and Indian Science Congress, invitation to conferences and academic institutions. But to Ramachandra the excitement of doing mathematics is its own reward. I wish him many years of happiness of creative pursuit.

I am very grateful to Prof. K. Srinivas for his help in preparing this foreword.

– R.P. Bambah

PREFACE

This volume represents the proceedings of the international conference *Analytic Number Theory* held at National Institute for Advanced Studies, Bangalore, during December 13 - 15, 2003. On this occasion Professor K. Ramachandra, who turned 70, was felicitated. We thank TIFR (Mumbai) and HRI (Allahabad) who have co-sponsored this conference. We thank the participants of the conference and the contributors to this volume and the referees. Thanks are also due to the administrative staff at NIAS for their help in organising the conference. Our special thanks are due to Mr. Ramakrishna Manja, Ms. M. Geetha, Ms. R. Indra and Dr. R. Thangadurai who took upon themselves various organisational matters and followed them up meticulously. We thank Prof. S.D. Adhikari for some editorial help. Finally we would like to thank Ramanujan Mathematical Society to consider our proceedings to publish under the DST-sponsored RMS Lecture Notes series in Mathematics. Publication of the present volume is co-sponsored by The Institute of Mathematical Sciences, Chennai.

August, 2006

R. Balasubramanian
K. Srinivas

Felicitation Messages

Let me congratulate you on 70 years of fascinating mathematics. As soon as I entered research, 30 years ago, yours became a familiar name; and your influence has remained with me ever since.
Time permits me to mention in detail only one strand of your work - but it is one that clearly demonstrates how important your research has been. A little over 20 years back you proved the first results on fractional moments of the Riemann Zeta-function. At first I could not believe they were correct!! Since then however the ideas have been extended in a number of ways. They have lead of course to a range of important new results about the Zeta-function and other Dirichlet series. But just as significantly the ideas have led to new conjectures on the moments of the Riemann Zeta-function. These conjectures provide the first successful test for the application of Random Matrix Theory in this area. Nowadays this is a growth area which has contributed much to our understanding of zeta-functions. And it can all be traced back to your work in the late 1970's.
So, let me congratulate you once again, and send you every good wish for the future.

— Roger Heath-Brown

I have great respect and affection for Professor Ramachandra who is the father of analytic number theory in the second half of the twentieth century in India.

— Alladi Krishnaswamy

At the celebration of your 70th birthday I congratulate you on your many excellent results in number theory, algebraic, transcendental and, above all, analytic. You have served the mathematical community not only by your research and teaching, but also by founding and editing for many years, together with your eminent student, R. Balasubramanian, The Hardy Ramanujan Journal, one of very few privately published mathematical journals. Let me wish you many more fruitful years.

— Andrzej Schinzel

Your colleagues from Department of Mathematics and Mechanics of Lomonosov Moscow State University warmly and heartly wish to express sincere congratulations on occasion of your seventieth birthday.
We know you as the great number-theoretist, the friend of mathematicians from Russia, and the nice person. Your mathematical school has the great influence on the development of the Number Theory in the whole world. In the name of our friendship, please, accept our sincere appreciation and gratitude.

— G.I. Arkhipov, V.N. Chubarikov, A.A. Karatsuba:
Y.V. Nesterenko, A.B. Shidlovski

It is a great pleasure to congratulate you on the occasion of your 70th birthday.
You have brought in the large contribution to development of number theory our favorite science. Your results are widely known to experts on number theory in many countries. Scientific contacts to you always have the large interest for us.
We wish you kind health, pleasures in life, successes in all your projects and new creative achievements.

— The chair of number theory of Moscow state university:
A.B. Shidlovskii, Yu. Nesterenko, A. Galochkin,
N. Moshevitin, V. Zudilin

I was very pleased to hear about the conference in honor of Professor Ramachandra on the occasion of his 70th birthday. I have great respect for Professor Ramachandra and have long admired his work. I think it is splendid that he should be honored in this way.

— Alan Baker

The post-Ramanujan period in India saw great advances, in several areas in modern mathematics. In analytic number theory, a beginning was made by my teacher Professor Anada Rau of Chennai. He was a student of G. H. Hardy and a friend and contemporary of Ramanujan at Cambridge. Others who followed him include T. Vijayaraghavan, S. S. Pillai and his student Sathe (both of whom had their career ended by early death) and V. Ganapathi Iyer — to name a few.
In the contemporary period in India, Professor K. Ramachandra occupies a special place in the field of analytic number theory. Besides doing outstanding research work on the many difficult problems involving the Riemann zeta function, he along with his talented students and coworkers built an enduring research group of analytic theorists in India which already enjoys a high reputation. I heard it said in some quarters that this group is one of the potential centers from where to expect a proof or disproof of the Riemann conjecture.
Ignoring temptations of moving to greener pastures outside his country, Ramachandra leads – like the wise men of yore – a life of simple living and high thinking. Like the famed late Sarvadaman Chowla, Ramachandra is full of modesty and humility. While self-importance reigns almost everywhere, he has no hesitation in declaring one of his students (Balu) as a greater mathematician than himself.
Let us all join in wishing Professor Ramachandra a long, healthy and mathematically fruitful life.

— M.V. Subbarao

I owe very much to you mathematically and personally, and in particular I gratefully remember my visit, through your initiative, to TIFR in 1985 which gave me a unique opportunity to mathematical work in a very pleasant and inspiring environment. Though I am afraid that because of other commitments I will be unable to participate your birthday conference, in any case I will be present in spirit and wish best success both to the conference and for you personally.

— Matti Jutila

I have been lucky to be associated with Professor Ramachandra starting from my student days and I wish him a very very long and happy life on this nice day.

— A. Sankaranarayanan

Doctoral Students of Professor K. Ramachandra

1. T.N. Shorey
2. S. Srinivasan
3. R. Balasubramanian
4. M. J. Narlikar
5. V.V. Rane
6. A. Sankaranarayanan
7. K. Srinivas

Table of contents

The Riemann Zeta Function and Related Themes – 2006, pp. 1–14

Zero-sum problems in combinatorial number theory

Sukumar Das Adhikari and Purusottam Rath

Dedicated to Professor K. Ramachandra on his 70th birthday

Abstract

In the present article, we survey an area consisting of some zero-sum problems in Combinatorial Number Theory, mention some open questions in the area, give references to some related questions and try to summarize some of the recent developments including the recent proof of a conjecture due to Kemnitz.

1 Introduction

A familiar high school exercise says that given any sequence of n integers $a_1, a_2, \ldots, a_n$, some of the a_i's sum up to $0 \pmod{n}$. In other words, $\exists$ nonempty $I \subset \{1, \ldots, n\}$ such that

$$\sum_{i \in I} a_i = 0 \pmod{n}. \tag{1}$$

Indeed, if one considers the sums $s_1 = a_1, s_2 = a_1 + a_2, \ldots, s_n = a_1 + \cdots + a_n$, then either an s_i is $0 \pmod{n}$ or at least two of the s_i's are equal modulo n.

Direct generalization of the above problem to arbitrary finite abelian groups is an important question. Though we do not take up this question in the present article, we shall mention this problem in Remark 6 and once again in the final section. We refer our readers to the articles of Olson [32] [33], Geroldinger and Schneider [24], Gao [18], Delorme, Ordaz and Quiroz [13], Thangadurai [23] and the references cited in these papers for further information related to that question.

We take up generalization of the problem in another direction, where one asks about prescribing the size of I (with some obvious constraints) in (1). In this particular direction, a theorem of Erdős, Ginzburg and Ziv [15] (henceforth, referred to as the EGZ theorem) says that

Theorem 1 (EGZ theorem). *For any positive integer n, any sequence $a_1, a_2, \ldots, a_{2n-1}$ of $2n-1$ integers has a subsequence of n elements whose sum is 0 modulo n.*

A prototype of zero-sum theorems, the EGZ theorem continues to play a central role in the development of this area of combinatorics. In the present article, we survey this area, give references to some related questions and try to summarize some of the recent developments including the recent proof of Kemnitz's conjecture.

Apart from the original paper of Erdős, Ginzburg and Ziv [15], there are many proofs of the above theorem available in the literature (see [1], [5], [7], [31] for instance). We shall present two of these in the next section. Another proof will be indicated in Remark 5.

The higher dimensional analogue of the EGZ theorem, which was considered initially by Harborth [26] and Kemnitz [28] has given rise to a very active area of combinatorics today. In Sections 3 and 4, we shall take up this theme and mention some results of Alon,

Dubiner [5] [6] and Rónyai [38] in this direction along with other related questions. We shall end Section 4 with a sketch of Reiher's recent proof of Kemnitz's conjecture, thus establishing the two dimensional analogue of the EGZ theorem.

Finally, in Section 5, we briefly describe the analogous questions related to general finite groups.

2 The EGZ theorem

Proof of Theorem 1. We observe that the essence of the EGZ theorem lies in the case when n is a prime. For the case $n = 1$, there is nothing to prove and let us assume the result is true in the case when n is a prime. Now, we proceed by induction on the number of prime factors (counted with multiplicity) of n. Therefore, if $n > 1$ is not a prime, we write $n = mp$ where p is prime and assume that the result is true for all integers with number of prime factors less than that of n.

By our assumption, each subsequence of $2p - 1$ members of the sequence a_1, a_2, $\dots, a_{2n-1}$ has a subsequence of p elements whose sum is 0 modulo p. From the original sequence we go on repeatedly omitting such subsequences of p elements having sum equal to 0 modulo p. Even after $2m - 2$ such sequences are omitted, we are left with $2pm - 1 - (2m - 2)p = 2p - 1$ elements and we can have at least one more subsequence of p elements with the property that sum of its elements is equal to 0 modulo p.

Thus we have found $2m - 1$ pairwise disjoint subsets $I_1, I_2, \dots, I_{2m-1}$ of $\{1, 2, \dots, 2mp - 1\}$ with $|I_i| = p$ and $\sum_{j \in I_i} a_j \equiv 0 \pmod{p}$ for each i. We now consider the sequence $b_1, b_2, \dots, b_{2m-1}$ where for $i \in \{1, 2, \dots, 2m - 1\}$, b_i is the integer $\frac{1}{p} \sum_{j \in I_i} a_j$.

By the induction hypothesis, this new sequence has a subsequence of m elements whose sum is divisible by m. The union of the corresponding sets I_i will supply the desired subsequence of $mp = n$ elements of the original sequence such that the sum of the elements of this subsequence is divisible by n. □

Let us now proceed to establish the result in the case $n = p$, a prime. For the first proof presented here, we shall need the following result (for a proof of which, one may look into [1] or [27], for instance).

Theorem 2.A. (Chevalley-Warning) *Let $f_i(x_1, x_2, \dots, x_n)$, $i = 1, \dots, r$, be r polynomials in $\mathbb{F}_q[x_1, x_2, \dots, x_n]$ such that the sum of the degrees of these polynomials is less than n and $f_i(0, 0, \dots, 0) = 0$, $i = 1, \dots, r$. Then there exists $(\alpha_1, \alpha_2, \dots, \alpha_n) \in \mathbb{F}_q^n$ with not all α_i's zero, which is a common solution to the system $f_i(x_1, x_2, \dots, x_n) = 0$, $i = 1, \dots, r$.*

Here and in what follows, for any prime power q, $\mathbb{F}_q$ will denote the finite field with q elements and the symbol $\mathbb{F}_q^*$ will denote the multiplicative group of non-zero elements of $\mathbb{F}_q$.

Now we proceed to prove Theorem 1 for the case $n = p$, a prime. Given a sequence $a_1, a_2, \dots, a_{2p-1}$ of elements in $\mathbb{F}_p$, we consider the following system of two polynomials in $(2p - 1)$ variables over the finite field $\mathbb{F}_p$:

$$\sum_{i=1}^{2p-1} a_i x_i^{p-1} = 0, \qquad \sum_{i=1}^{2p-1} x_i^{p-1} = 0.$$

Since $2(p - 1) < 2p - 1$ and $x_1 = x_2 = \cdots = x_{2p-1} = 0$ is a solution, by Theorem 2.A above, there is a nontrivial solution $(y_1, \dots, y_{2p-1})$ of the above system. By Fermat's little theorem, writing $I = \{i : y_i \neq 0\}$, from the first equation it follows that $\sum_{i\in I} a_i = 0$ and from the second equation we have $|I| = p$.

For our second proof of the 'prime case' of EGZ theorem, we shall need the following generalized version of *Cauchy-Davenport inequality* ([11], [12], can also look into [30] or [31] for instance):

Theorem 2.B. (Cauchy-Davenport) *Let $A_1, A_2, \ldots, A_h$ be non-empty subsets of $\mathbb{F}_p$. Then*

$$|\sum_{i=1}^{h} A_i| \geq min\left(p, \sum_{i=1}^{h} |A_i| - h + 1\right).$$

(Here $\sum_{i=1}^{h} A_i$ is the set consisting of all elements of $\mathbb{F}_p$ of the form $\sum_{i=1}^{h} a_i$ where $a_i \in A_i$.)

Now, for a prime p, we consider representatives modulo p in the interval $0 \leq a_i \leq p-1$ for the given elements and rearranging, if necessary, we assume that

$$0 \leq a_1 \leq a_2 \leq \cdots \leq a_{2p-1} \leq p-1.$$

We can now assume that

$$a_j \neq a_{j+p-1}, \quad \text{for } j = 1, \ldots, p-1.$$

For otherwise, the p elements $a_j, a_{j+1}, \ldots, a_{j+p-1}$ being equal, the result holds trivially.

Now, applying Theorem 2.B on the sets

$$A_j := \{a_j, a_{j+p-1}\}, \quad \text{for } j = 1, \ldots, p-1,$$

so that

$$\left|\sum_{i=1}^{p-1} A_i\right| \geq min\left(p, \sum_{i=1}^{p-1} |A_i| - (p-1) + 1\right) = p,$$

we have

$$-a_{2p-1} \in \sum_{i=1}^{p-1} A_i$$

and hence once again we have established EGZ theorem for the case when n is a prime. □

Remark 1. The EGZ theorem as well as many other zero-sum results can also find their place in a larger class of results in combinatorics. More precisely, a result saying that a substructure can not avoid certain regularity properties of the original structure because the 'size' of the substructure is 'large' enough, or a structure which sufficiently 'big' has certain unavoidable regularities, is termed as a *Ramsey-type theorem* in combinatorics.

Remark 2. Let us now observe that in Theorem 1, the number $2n-1$ is the smallest positive integer for which the theorem holds. In other words, if $f(n)$ denotes the smallest positive integer such that given a sequence $a_1, a_2, \ldots, a_{f(n)}$ of not necessarily distinct integers, there exists a set $I \subset \{1, 2, \ldots, f(n)\}$ with $|I| = n$ such that $\sum_{i\in I} a_i \equiv 0 \pmod{n}$, then $f(n) = 2n-1$. This can be seen as follows. From Theorem 1, it follows that $f(n) \leq 2n-1$. On the other hand, if we take a sequence of $2n-2$ integers such that $n-1$ among them are 0 modulo n and the remaining $n-1$ are 1 modulo n, then clearly, we do not have any subsequence of n elements sum of whose elements is 0 modulo n.

The following interesting result of Bollobás and Leader [7] clearly implies Theorem 1 by taking $r = n-1$.

Theorem 2. *Let G be an abelian group of order n and r be a positive integer. Let A denote the sequence* $a_1, a_2, \ldots, a_{n+r}$ *of* $n + r$ *not necessarily distinct elements of G. Then if 0 is not an n-sum, the number of distinct n-sums of A (by an n-sum one means a sum* $a_{i_1} + \cdots + a_{i_n}$ *of a subsequence of length n of A) is at least* $r + 1$.

A simple proof of the above theorem has been given by Yu [45] using the following result of Scherk [39] (see also [29] and Theorem 15′ of Chap. 1 in [25]), which is an analogue of Cauchy-Davenport inequality.

Lemma 1. *Let B and C be two subsets of an abelian group G of order n. Suppose* $0 \in B \cap C$ *and suppose that the only solution of* $b + c = 0$, $b \in B$, $c \in C$ *is* $b = c = 0$. *Then,*

$$|B + C| \geq |B| + |C| - 1.$$

Remark 3. Similar to the notation used in Theorem 2.B, here $B + C$ is the set consisting of all elements of G of the form $b + c$ where $b \in B$ and $c \in C$ and it is not difficult to check that with the assumption in the lemma, the number $|B| + |C| - 1$ cannot exceed $|G|$.

3 Higher dimensional analogue

As in Remark 2, for any positive integer d we define $f(n, d)$ to be the smallest positive integer such that given a sequence of $f(n, d)$ number of not necessarily distinct elements of $\mathbb{Z}^d$, there exists a subsequence $x_{i_1}, x_{i_2}, \ldots, x_{i_n}$ of length n such that its centroid $(x_{i_1} + x_{i_2} + \cdots + x_{i_n})/n$ also belongs to $\mathbb{Z}^d$. In other words, $f(n, d)$ is the smallest positive integer N such that every sequence of N elements in $(\mathbb{Z}/n\mathbb{Z})^d$ has a subsequence of n elements which add up to $(\underbrace{0, 0, \ldots, 0}_{d \text{ times}})$. We observe that $f(n, 1) = f(n)$ where $f(n)$ is as defined in Remark 2.

This higher dimensional analogue was first considered by Harborth [26]; he observed the following general bounds for $f(n, d)$.

Since the number of elements of $(\mathbb{Z}/n\mathbb{Z})^d$ having coordinates 0 or 1 is 2^d, considering a sequence where each of these elements are repeated $(n - 1)$ times, one observes that

$$1 + 2^d(n - 1) \leq f(n, d). \tag{2}$$

Again, observing that in any sequence of $1 + n^d(n - 1)$ elements of $(\mathbb{Z}/n\mathbb{Z})^d$ there will be at least one vector appearing at least n times, we have

$$f(n, d) \leq 1 + n^d(n - 1). \tag{3}$$

For $d = 1$, the EGZ theorem gives the exact value

$$f(n, 1) = 2n - 1.$$

For the case $d = 2$ also, the lower bound in (2) is expected to give the right magnitude of $f(n, 2)$ and this expectation, which is known as *Kemnitz Conjecture* in the literature, has been recently established by Reiher. We shall discuss Reiher's result in the next section. Historically, the first result in this direction was contained in the above mentioned paper of Harborth [26] where he proved that $f(3, 2) = 9$. Kemnitz [28] established this conjecture when n is of the form $2^e 3^f 5^g 7^h$. However, the lower bound given in (2) is known not to be tight in general. Harborth [26] proved that $f(3,3) = 19$; this is strictly greater than the lower bound 17 which one obtains from (2). Different proofs of the result $f(3,3) = 19$ appeared since then (see [8] and [3] for instance; see also [6] for some more references in this regard). However, Harborth's result on $f(3,3)$ did not rule out the possibility that for a fixed dimension d, for a sufficiently large prime p the lower bound in (2) might determine the exact value for $f(p,d)$. But a recent result of Elsholtz [14] in this direction, rules out such possibilities. We shall come back to this theme very shortly.

Another important observation made by Harborth [26] was the following:

$$f(mn,d) \leq \min(f(n,d) + n(f(m,d) - 1), f(m,d) + m(f(n,d) - 1)). \tag{4}$$

This result follows by an elementary argument of the same nature as was adopted in deriving Theorem 1 from the result in the 'prime case'.

Harborth [26] observed that from (2), (3) and (4), one can easily derive the exact value for $f(2^e,d)$ for any $d \geq 2$. More precisely, for $n = 2$ the lower and upper bounds for $f(2,d)$, given respectively by (2) and (3), are both $2^d + 1$ and assuming $f(2^r,d) = (2^r - 1)2^d + 1$, $f(2^s,d) = (2^s - 1)2^d + 1$ for some particular d, by (4) it follows that $f(2^{r+s},d) = (2^{r+s} - 1)2^d + 1$.

However, for all odd primes p and $d \geq 3$, we have a long way to go regarding the exact values for $f(p,d)$.

Coming back to the cases $d \geq 3$, the lower bound in (2) is known not to be the exact value of $f(p,d)$ for all odd primes p. As mentioned earlier, a particular instance of this phenomenon was observed by Harborth [26] by proving that $f(3,3) = 19$. The following general result in this direction was proved by Elsholtz [14].

Theorem 3. *For an odd integer $n \geq 3$, the following inequality holds:*

$$f(n,d) \geq \left(\frac{9}{8}\right)^{[\frac{d}{3}]} (n-1)2^d + 1.$$

Thus the lower bound in (2) is not the correct value of $f(n,d)$ for $d \geq 3$.

Now, one observes that the gap is quite large between the lower and the upper bounds given respectively in (2) and (3). A very important result of Alon and Dubiner [6] says that the growth of $f(n,d)$ is linear in n; when d is fixed and n is increasing, this is much better as compared to the upper bound given by (3). More precisely, Alon and Dubiner [6] prove the following.

Theorem 4. *There is an absolute constant $c > 0$ so that for all n,*

$$f(n,d) \leq (cd\log_2 d)^d n.$$

The proof of Theorem 4 due to Alon and Dubiner combines techniques from additive number theory with results about the expansion properties of Cayley graphs with given eigenvalues. In the same paper [6] the authors conjecture that the estimate in Theorem 4 can possibly be improved. More precisely, the existence of an absolute constant c is predicted such that

$$f(n,d) \leq c^d n, \text{ for all } n \text{ and } d.$$

4 The two dimensional case

As have been mentioned, with Reiher's recent proof of Kemnitz's conjecture the problem has been solved in the two dimensional case. We go through the historical development to some extent.

In the two dimensional case, in a very significant paper [5], Alon and Dubiner proved that

Theorem 5. *We have*

$$f(n, 2) \leq 6n - 5.$$

One can observe that by an argument similar to the one used in the proof of Theorem 1, the inequality $f(p, 2) \leq 6p - 5$, for every prime p, implies $f(n, 2) \leq 6n - 5$, for every n. The proof of the fact that $f(p, 2) \leq 6p - 5$, as given in this paper of Alon and Dubiner, is ingenious and uses algebraic tools such as the theorem of Chevalley and Warning (Theorem 2.A) and the algebra of permanents. It also uses the EGZ theorem, the result in the one dimensional case. It has been indicated in [5] that the proof can be modified to yield the stronger result that $f(p, 2) \leq 5p - 2$. A relatively simple proof of a slightly weaker version of Theorem 5 is also sketched in this paper.

We now state a sharper result due to Rónyai [38].

Theorem 6. *For a prime p, we have*

$$f(p, 2) \leq 4p - 2.$$

Remark 4. As we have mentioned before, from the inequality $f(p, 2) \leq 6p - 5$, for every prime p, by an argument similar to the one used in the proof of Theorem 1, one gets the result $f(n, 2) \leq 6n - 5$, for every n. Such would be the case for the bound $f(n, 2) \leq 4n - 3$ of Kemnitz's conjecture. However, this argument does not go through for the bound given by the above theorem. But, as mentioned in Rónyai [38], it is not difficult to observe that Theorem 6 along with (4) implies that

$$f(n, 2) \leq \frac{41}{10}n.$$

Gao [19] obtained the following generalization of the result (Theorem 6) of Rónyai [38] mentioned before.

Theorem 7. *For an odd prime p and a positive integer r, we have*

$$f(p^r, 2) \leq 4p^r - 2.$$

Following Gao [19], we now sketch a proof of Theorem 7. We note that this proof proceeds along a line which is quite different from the proof of Theorem 6 as given by Rónyai [38].

The proof uses the following special case of a very elegant result of Olson [32]. Apart from being interesting on its own right, it has several important results as its immediate corollaries.

Lemma 2 (Olson). *For a prime p, let $s_1, s_2, \ldots, s_k$ be a sequence S of elements in $(\mathbb{Z}/p^r\mathbb{Z})^d$ such that $k \geq 1 + d(p^r - 1)$. Then, writing $f_e(S)$ for the number of subsequences of even* *length of S which sum up to zero and $f_o(S)$ for the number of subsequences of odd length which sum up to zero, we have*

$$f_e(S) - f_o(S) \equiv -1 \pmod{p}.$$

First we note the following two corollaries to Lemma 2; these will be used to prove Theorem 7. Later we shall remark about few more consequences of the above lemma of Olson.

Lemma 3. *If S is a zero-sum sequence of $3p^r$ elements in $(\mathbb{Z}/p^r\mathbb{Z})^2$, then S contains a zero-sum subsequence of length p^r.*

For any sequence S of elements in $(\mathbb{Z}/p^r\mathbb{Z})^2$, if $r(S)$ denotes the number of zero-sum subsequences W of S with $|W| = 2p^r$, one has the following.

Lemma 4. *Let T be a sequence of elements in $(\mathbb{Z}/p^r\mathbb{Z})^2$ with $3p^r - 2 \leq |T| \leq 4p^r - 1$. Suppose that T contains no zero-sum subsequence of length p^r. Then*

$$r(T) \equiv -1 \pmod{p}.$$

Both the above lemmas follow easily from Lemma 2 by appending 1 as the third coordinate to each of the elements in S and T respectively. For instance, if S in Lemma 3 is $(a_1, b_1), (a_2, b_2), \ldots, (a_m, b_m)$, where $m = 3p^r$, one considers the sequence $S' = (a_1, b_1, 1), (a_2, b_2, 1), \ldots, (a_l, b_l, 1)$ with $l = 3p^r - 2$. Now, by Lemma 2, S' and hence $S_1 = (a_1, b_1), (a_2, b_2), \ldots, (a_l, b_l)$ must have a zero-sum subsequence, length of which must be p^r or $2p^r$. If there is a zero-sum subsequence S_2 of S_1 of length $2p^r$, then its complement in S provides us with one such with length p^r.

Proof of Theorem 7. If possible, suppose that there is a sequence S of elements in $(\mathbb{Z}/p^r\mathbb{Z})^2$ such that S is of length $4p^r - 2$ and S has no zero-sum subsequence of length p^r.

By Lemma 4,

$$r(T) \equiv -1 \pmod{p},$$

for every subsequence T of S with $|T| \geq 3p^r - 2$.

We have

$$\sum_{T \subset S, |T| = 3p^r - 2} r(T) = \binom{4p^r - 2 - 2p^r}{3p^r - 2 - 2p^r} r(S).$$

Hence

$$\sum_{T \subset S, |T| = 3p^r - 2} (-1) \equiv \binom{2p^r - 2}{p^r - 2} (-1) \pmod{p}.$$

Thus

$$\binom{4p^r - 2}{3p^r - 2} \equiv \binom{2p^r - 2}{p^r - 2} \pmod{p},$$

which would imply that

$$3 \equiv \binom{4p^r - 2}{3p^r - 2} \equiv \binom{2p^r - 2}{p^r - 2} \equiv 1 \pmod{p}$$

- a contradiction. □

Remark 5. As was observed by Alon and Dubiner [5], the EGZ theorem follows almost immediately from Lemma 5. More precisely, for any prime p, given any sequence $a_1, a_2, \dots, a_{2p-1}$ of elements in $(\mathbb{Z}/p\mathbb{Z})$, we just consider the sequence

$$(a_1, 1), (a_2, 1), \dots, (a_{2p-1}, 1)$$

in $(\mathbb{Z}/p\mathbb{Z})^2$.

Remark 6. Regarding implications of Lemma 2 we must mention that in the original paper of Olson [32], the lemma was used to find the value of Davenport's constant $D(G)$ for a finite abelian p-group G. For any finite abelian group G, the Davenport's constant $D(G)$ is an important combinatorial invariant defined to be the smallest positive integer s such that for any sequence $g_1, g_2, \dots, g_s$ of (not necessarily distinct) elements of G, there is a nonempty $I \subset \{1, \dots, s\}$ such that $\sum_{\in I} g_i = 0$. For relations between Kemnitz's conjecture and a conjecture involving the Davenport's constant and some other conjectures related to zero-sum problems, one may look into some papers of Thangadurai [43] and Gao and Geroldinger [21]. In Section 5, we shall have an occasion to state an important relation (due to Gao [16]) between the Davenport's constant and another constant emerging out from a natural generalization of the EGZ theorem for finite abelian groups.

We now consider a generalization of $f(n, d)$ as defined in the beginning of Section 3. Let $f_r(n, d)$ denote the smallest positive integer such that given any sequence of $f_r(n, d)$ elements in $(\mathbb{Z}/n\mathbb{Z})^d$, there exists a subsequence of (rn) elements whose sum is zero in $(\mathbb{Z}/n\mathbb{Z})^d$. Thus $f_1(n, d) = f(n, d)$.

As has been mentioned in a recent paper of Gao and Thangadurai [22], one can derive (as in [21]) that

$$f_r(n, 2) = (r + 2)n - 2, \text{ for integers } r \geq 2 \tag{5}$$

from the known results about the Davenport's constant for finite abelian p-groups and by using Reiher's result on the exact value of $f_1(n, 2)$.

The particular cases of $f_r(p, 2)$ for $r \geq 2$ where p is a prime had been established in [2] by different arguments. Exact values of $f_r(n, 1)$ for $r \geq 1$ can be easily obtained from the EGZ theorem. We remark that at the time when the paper [2] had been written, Reiher's result was yet to be established.

As had been mentioned in the introduction, we shall conclude this section with a sketch of Reiher's recent proof of Kemnitz's conjecture. We mention that some interesting partial results towards the conjecture of Kemnitz and some related results had been obtained by Gao [17], Thangadurai [42] and Sury and Thangadurai [41].

The present version of the proofs of the equations (6), (8), (9) and (10), namely the idea of deriving them from Lemma 2, is different from that in the original proof of Reiher.

Theorem 8 (Reiher). *For an odd prime p, we have*

$$f(p, 2) = 4p - 3.$$

Proof: By (2) it suffices to show that $f(p, 2) \leq 4p - 3$. Let X be a sequence of $4p - 3$ elements of $(\mathbb{Z}/p\mathbb{Z})^2$ with no zero-sum subsequence of length p.

For any subsequence J of X, let (n, J) denote the number of zero-sum subsequences of J of length n.

In particular, we have $(p, X) = 0$ and hence by a result of Alon and Dubiner [5], $(3p, X) = 0$.

First, we deduce the following consequence of Lemma 2.

For any subsequence $J = (a_1, b_1), (a_2, b_2), \ldots, (a_{(3p-3)}, b_{(3p-3)})$ of X, we have

$$1 - (p-1, J) + (2p-1, J) + (2p, J) \equiv 0 \pmod{p}. \tag{6}$$

The proof of the statement (6) is as follows. We consider the sequence J_1 consisting of the $3p-3$ elements in $(\mathbb{Z}/p\mathbb{Z})^3$ given by $(a_1, b_1, 1), \ldots, (a_{3p-3}, b_{3p-3}, 1)$ along with the element $(0, 0, 1)$. Now the length of J_1 is $3p - 2 \geq 1 + 3(p-1)$, hence we can apply Lemma 2 of Olson. Since the third co-ordinate of all the above elements is 1, any subsequence of even length of J_1 which sums up to zero has to be of length $2p$ while any subsequence of odd length of J_1 which sums up to zero has to be of length p. Any subsequence of length $2p$ of J_1 which sums up to zero and which does not involve the last term $(0, 0, 1)$ gives a subsequence of length $2p$ of J which sums up to zero and vice versa. However any subsequence of length $2p$ of J_1 which sums up to zero and which involves the last term $(0, 0, 1)$ actually gives a subsequence of length $2p-1$ of J which sums up to zero and vice versa. Similar is the case for the zero-sum subsequences of length p. Now using the above result of Olson and keeping in mind the fact that $(p, J) = 0$, we get the desired result.

Now,

$$\sum [1 - (p-1, I) + (2p-1, I) + (2p, I)] \equiv 0 \pmod{p}.$$

where I runs over all possible subsequences I of length $3p-3$ of X.

Therefore,

$$\begin{aligned}\binom{4p-3}{3p-3} - \binom{3p-2}{2p-2}(p-1, X) + \binom{2p-2}{p-2}(2p-1, X)\\ + \binom{2p-3}{p-3}(2p, X) \equiv 0 \pmod{p},\end{aligned}$$

which gives

$$3 - 2(p-1, X) + (2p-1, X) + (2p, X) \equiv 0 \pmod{p}. \tag{7}$$

Further,

$$(2p, J) \equiv -1 \pmod{p}, \tag{8}$$

for any subsequence J of X of length $3p-2$ or $3p-1$, and

$$(2p, X) \equiv -1 \pmod{p}, \tag{9}$$

by considering the relevant sequences in $(\mathbb{Z}/p\mathbb{Z})^3$ obtained by putting 1 in the last coordinate and using Lemma 2.

Now, once again following the argument employed in deducing (6) and using (9), we have

$$(p-1, X) - (2p-1, X) + (3p-1, X) \equiv 0 \pmod{p}. \tag{10}$$

Finally, we prove that

$$(p-1, X) \equiv (3p-1, X) \pmod{p}. \tag{11}$$

We proceed as follows. Let θ denote the number of partitions of $X = A \cup B \cup C$ into disjoint subsequences A, B and C where A is a zero-sum subsequence of length $p-1$ and C is a zero-sum subsequence of length $2p$. We count θ in two different ways.

First,

$$\theta = \sum_A (2p, X \setminus A),$$

where the summation runs over all zero-sum subsequences of length $p - 1$.

Using (8), this gives

$$\theta \equiv -(p-1, X) \pmod{p}. \tag{12}$$

Again,

$$\theta = \sum_{A'} (2p, A'),$$

where the summation runs over all zero-sum subsequences of length $3p - 1$ and by using (8), we have

$$\theta \equiv -(3p-1, X) \pmod{p}. \tag{13}$$

From (12) and (13), we get (11).

Now, adding (7) and (10), by (11), we have

$$(2p, X) \equiv -3 \pmod{p},$$

–contradicting (9). □

Since the case $p = 2$ had been already established, by Remark 4 the above result really settles Kemnitz's conjecture, that is $f(n, 2) = 4n - 3$, for any positive integer n.

5 EGZ for finite groups

One observes that by taking $r = n - 1$ in Theorem 2 of Bollobás and Leader [7], not only we get Theorem 1, but also the following natural generalization of the EGZ theorem for finite abelian groups.

Theorem 9. *Let G be an abelian group of order n. Then given any sequence $g_1, g_2, \ldots, g_{2n-1}$ of $2n-1$ elements of G, there exists a subsequence of n elements whose sum is the identity element 0 of G.*

However, it is not difficult to see that following the method employed in deriving Theorem 1 from the 'prime case', and appealing to the structure theorem for finite abelian groups, one can derive Theorem 9 from the EGZ theorem.

We note that for the cyclic group of order n, $2n - 1$ is the smallest number satisfying the above property. That is, if for any abelian group G of order n, if $\mathrm{ZS}(G)$ is the smallest integer t such that for any sequence of t elements of G, there exists a subsequence of n elements whose sum is the identity element 0 of G, then we have

$$\mathrm{ZS}(\mathbb{Z}/n\mathbb{Z}) = 2n - 1.$$

By Theorem 9, $\mathrm{ZS}(G) \leq 2n - 1$, for any abelian group G of order n. However, for a non-cyclic abelian group G of order n, $\mathrm{ZS}(G)$ need not be equal to $2n - 1$. In this direction, a result of Alon, Bialostocki and Caro [4] says that for a non-cyclic abelian group G of order n, $\mathrm{ZS}(G) \leq \frac{3n}{2}$ and the bound $\frac{3n}{2}$ is realized only by groups of the form $\mathbb{Z}/2\mathbb{Z} \oplus \mathbb{Z}/2m\mathbb{Z}$. Subsequently Caro [9] showed that if a non-cyclic abelian group G of order n is not of the form $\mathbb{Z}/2\mathbb{Z} \oplus \mathbb{Z}/2m\mathbb{Z}$, then $\mathrm{ZS}(G) \leq \frac{4n}{3} + 1$ and this bound is realized only by groups of the form $\mathbb{Z}/3\mathbb{Z} \oplus \mathbb{Z}/3m\mathbb{Z}$. Further generalization of the same nature have been obtained by Ordaz and Quiroz [36] rather recently stating that apart from the groups $\mathbb{Z}/2\mathbb{Z} \oplus \mathbb{Z}/2m\mathbb{Z}$ and $\mathbb{Z}/3\mathbb{Z} \oplus \mathbb{Z}/3m\mathbb{Z}$ which appear in the last statement, for any non-cyclic abelian group G of order n, $\mathrm{ZS}(G) \leq \frac{5n}{4} + 2$ and equality holds only for groups of the form $\mathbb{Z}/4\mathbb{Z} \oplus \mathbb{Z}/4m\mathbb{Z}$. Further generalization of these results describing the situation with abelian groups G having smaller values of $\mathrm{ZS}(G)$ may involve groups other than $\mathbb{Z}/2\mathbb{Z} \oplus \mathbb{Z}/2m\mathbb{Z}$, $\mathbb{Z}/3\mathbb{Z} \oplus \mathbb{Z}/3m\mathbb{Z}, \ldots, \mathbb{Z}/r\mathbb{Z} \oplus \mathbb{Z}/rm\mathbb{Z}$, for positive integers r. One should mention that the method of Ordaz and Quiroz [36] involves obtaining an upper bound for the Davenport's constant $D(G)$ (as defined in Remark 6) of the relevant groups G and using the following beautiful result of Gao [16] which links $\mathrm{ZS}(G)$ with $D(G)$.

Theorem 10. *If G is a finite abelian group of order n, then* $\mathrm{ZS}(G) = D(G) + n - 1$.

One would also like to know the validity of the statement of Theorem 9 for general finite groups.

For a finite solvable group G, by induction on the length k of a minimal abelian tower $(0) = G_k \subset G_{k-1} \subset \cdots \subset G_0 = G$ and using the result in Theorem 9, one can easily derive [35] (see also [40]) the following result employing the same argument as employed in the proof of Theorem 1 in deriving the general case from the 'prime case'.

Theorem 11. *Let G be a finite solvable group (written additively) of order n. Then given any sequence $g_1, g_2, \ldots, g_{2n-1}$ of $2n - 1$ elements of G, there exist n distinct indices $i_1, \ldots, i_n$ such that*

$$g_{i_1} + g_{i_2} + \cdots + g_{i_n} = 0.$$

The above result holds without the assumption that the group G is solvable. This follows from the following general result of Olson [34].

Theorem 12. *Let G be a finite group (written additively) of order $n > 1$. Let S be a sequence $g_1, g_2, \ldots, g_{2n-1}$ of elements of G in which no element appears more than n times. Then G has a subgroup K of order $k > 1$ and S has a subsequence $T = a_1, \ldots, a_{n+k-1}$ such that*

i) *K is a normal subgroup of the subgroup H of G generated by $\{a_1, \ldots, a_{n+k-1}\}$,*

ii) *there exists an $a \in H$ such that $a_i \in a + K = K + a$ for $1 \leq i \leq n + k - 1$, and*

iii) *K is the set of all sums $a_{i_1} + \cdots + a_{i_n}$ where $i_1, \ldots, i_n$ are n distinct indices (in any order) in $\{1, \ldots, n + k - 1\}$.*

We note that for a non-abelian group G of order n, given a sequence of $2n-1$ elements of G, we are not ensured that there is a subsequence of n elements which adds up to the identity, rather a permutation of a subsequence of length n will do so. However, it is conjectured [34] that there must be a subsequence of n elements which adds up to the identity. This is not known even for solvable groups.

References

[1] Sukumar Das Adhikari, *Aspects of combinatorics and combinatorial number theory*, Narosa, New Delhi, 2002.

[2] Sukumar Das Adhikari and Purusottam Rath, Remarks on some zero-sum problems, *Expo. Math.* **21**, no. 2, 185–191 (2003).

[3] Sukumar Das Adhikari, Stephan Baier and Purusottam Rath, An extremal problem in lattice point combinatorics, Preprint.

[4] N. Alon, A. Bialostocki and Y. Caro, Extremal zero-sum problems, manuscript.

[5] N. Alon and M. Dubiner, Zero-sum sets of prescribed size. Combinatorics, Paul Erdős is eighty (Volume 1), Keszthely (Hungary), 33–50 (1993).

[6] N. Alon and M. Dubiner, A lattice point problem and additive number theory, *Combinatorica* **15**, 301–309 (1995).

[7] Béla Bollobás and Imre Leader, The number of k-sums modulo k, *J. Number Theory*, **78**, no. 1, 27–35 (1999).

[8] J.L. Brenner, Problem 6298, *Amer. Math. Monthly*, **89**, 279–280 (1982).

[9] Yair Caro, Zero-sum subsequences in abelian non-cyclic groups, *Israel Journal of Mathematics*, **92**, 221–233 (1995).

[10] Yair Caro, Zero-sum problems – A survey, *Discrete Math.* **152**, 93–113 (1996).

[11] A.L. Cauchy, Recherches sur les nombres, *J. École polytech.*, **9**, 99–116, (1813).

[12] H. Davenport, On the addition of residue classes, *J. London Math. Soc.*, **10**, 30–32, (1935).

[13] C. Delorme, O. Ordaz and D. Quiroz, Some remarks on Davenport constant, *Discrete Math.*, **237**, 119–128 (2001).

[14] Christian Elsholtz, Lower bounds for multidimensional zero sums, *Combinatorica*, **24**, no. 3, 351–358 (2004).

[15] P. Erdős, A. Ginzburg and A. Ziv, Theorem in the additive number theory, *Bull. Research Council Israel*, **10F**, 41–43 (1961).

[16] W.D. Gao, A combinatorial problem on finite abelian groups, *J. Number Theory*, **58**, 100–103 (1996).

[17] W.D. Gao, On zero-sum subsequences of restricted size, *J. Number Theory*, **61**, 97–102 (1996).

[18] W.D. Gao, On Davenport's constant of finite abelian groups with rank three, *Discrete Math.*, **222**, 111–124 (2000).

[19] W.D. Gao, A note on a zero-sum problem, *J. Combinatorial Theory*, Ser. A, **95** no. 2, 387–389 (2001).

[20] W.D. Gao, On zero-sum subsequences of restricted size II, *Discrete Math.*, **271**, 51–59 (2003).

[21] W.D. Gao and A. Geroldinger, On zero-sum sequences in $\mathbb{Z}/n\mathbb{Z} \oplus \mathbb{Z}/n\mathbb{Z}$, *Integers: Electronic Journal of Combinatorial Number Theory*, **3**, #A8, 45 pp. (2003).

[22] W.D. Gao and R. Thangadurai, On zero-sum sequences of prescribed length, *Aequationes Math.*, to appear.

[23] W. D. Gao and R. Thangadurai, Davenport constant and non-abelian version of Erdős-Ginzburg-Ziv theorem, to appear.

[24] A. Geroldinger and R. Schneider, On Davenport's constant, *J. Combinatorial Theory*, Ser. A, **61** no. 2, 147–152 (1992).

[25] H. Halberstam and K.F. Roth, *Sequences*, 2nd Ed., Springer, 1983.

[26] H. Harborth, Ein Extremalproblem für Gitterpunkte, *J. Reine Angew. Math.* **262/263**, 356–360 (1973).

[27] Kenneth Ireland and Michael Rosen, *A Classical Introduction to Modern Number Theory*, 2nd edition, Springer-Verlag, (1990).

[28] A. Kemnitz, On a lattice point problem, *Ars Combin.*, **16b**, 151–160 (1983).

[29] J.H.B. Kemperman and P. Scherk, Complexes in abelian groups, *Can. J. Math.*, **6**, 230–237 (1954).

[30] H.B. Mann, *Addition Theorems in Group Theory and Number Theory*, R.E. Krieger Publishing Company, Huntington, New York, 1976.

[31] Melvyn B. Nathanson, *Additive Number Theory: Inverse Problems and the Geometry of Sumsets*, Springer, 1996.

[32] J.E. Olson, On a combinatorial problem of finite Abelian groups, I, *J. Number Theory*, **1** 8–10 (1969).

[33] J.E. Olson, On a combinatorial problem of finite Abelian groups, II, *J. Number Theory*, **1** 195–199 (1969).

[34] J.E. Olson, On a combinatorial problem of Erdös, Ginzburg and Ziv, *J. Number Theory*, **8**, 52–57 (1976).

[35] J.E. Olson, An addition theorem for finite abelian groups, *J. Number Theory*, **9**, 63–70 (1977).

[36] O. Ordaz and D. Quiroz, The Erdős-Ginzburg-Ziv theorem in abelian non-cyclic groups, *Divulg. Mat.* **8** (2), 113–119 (2000).

[37] Christian Reiher, On Kemnitz's conjecture concerning lattice points in the plane, *Ramanujan J.*, to appear.

[38] L. Rónyai, On a conjecture of Kemnitz, *Combinatorica*, **20** (4), 569–573 (2000).

[39] P. Scherk, Solution to Problem 4466, *Amer. Math. Monthly*, **62**, 46–47 (1955).

[40] B. Sury, The Chevalley-Warning theorem and a combinatorial question on finite groups, *Proc. Amer. Math. Soc.*, **127** (4), 951–953 (1999).

[41] B. Sury and R. Thangadurai, Gao's conjecture on zero-sum sequences, *Proc. Indian Acad. Sci.* (math. Sci.), **112** (3), 399–414 (2002).

[42] R. Thangadurai, On a conjecture of Kemnitz, *C. R. Math. Rep. Acad. Sci.* Canada, **23** (2), 39–45 (2001).

[43] R. Thangadurai, Interplay between four conjectures on certain zero-sum problems, *Expo. Math.* **20**, no. 3, 215–228 (2002).

[44] P. van Emde Boas, A combinatorial problem on finite Abelian groups II, *Z. W. (1969-007) Math. Centrum-Amsterdam.*

[45] Hong Bing Yu, A simple proof of a theorem of Bollobás and Leader, *Proc. American Math. Soc.*, **131**, no. 9, 2639–2640 (2003).

Sukumar Das Adhikari[1] and Purusottam Rath[2]
Harish-Chandra Research Institute
(Formerly, Mehta Research Institute)
Chhatnag Road, Jhusi,
Allahabad 211 019, India.

e-mail: [1]adhikari@mri.ernet.in,
[2]rath@mri.ernet.in

The Riemann Zeta Function and Related Themes – 2006, pp. 15–17

A generalization of Agarwal's theorem

Chandrashekar Adiga, N. Anitha and Jung Hun Han

Dedicated to Professor K. Ramachandra on his 70th birthday

Abstract

In this paper, we introduce the notion of N-parity partitions which generalizes the notion of odd-even partitions and thereby generalizes Agarwal's [1] result which is an analogue of the Rogers-Ramanujan's second partition identity.

1 Introduction

MacMahon [3] gave the folowing partition theoretic interpretation of Rogers-Ramanujan identities:

Theorem 1. *The number of partitions of n into parts with minimal difference 2 equals the number of partitions of n into parts which are congruent to* $\pm 1 \pmod 5$.

Theorem 2. *The number of partitions of n with minimal part 2 and minimal difference 2 equals the number of partitions of n into parts which are congruent to* $\pm 2 \pmod 5$.

Recently, A.K. Agarwal [1] proved a result which is an analogue of Theorem 2 for odd-even partitions using ordered partitions (compositions), In [2] Bhargava.S, Chandrashekar Adiga and N. Anitha obtained a generalization of Agarwal's result.

In this paper we introduce the notion of N-parity partitions which generalizes the notion of odd-even partitions and we thereby generalize Agarwal's [1] as well as the result of S. Bhargava, Chandrashekar Adiga and Anitha [2]. We also provide an alternative proof (Section 3) of our main result by following MacMahon's analysis.

2 Main Result

Given a positive integer N, we call a partition $\lambda = (\lambda_1, \lambda_2, \ldots, \lambda_m)$ an "N-parity partition" if $\lambda_i \equiv i \pmod{\mathrm{N}}, i = 1, 2, \ldots, m$. Let $C^{(l,N)}(v)$ denote the number of N-parity partition with largest part v, least part $\geq l$ and the difference between i^{th} and $(i-1)^{th}$ parts $\geq k_i$, $i = 2, 3, \ldots, m$ where each $k_i > 0$ and each $k_i \equiv 1 \pmod{\mathrm{N}}$. Let $C_m^{(l,N)}(v)$ denote the number of partitions enumerated by $C^{(l,N)}(v)$ into exactly m parts and let $C_m^{(l,N)}(v, n)$ denote the number of partitions of n among these. Let $D^{(l,N)}(v)$ denote the number of compositions of v into parts congruent to $1 \pmod{\mathrm{N}}$ with first part $\geq l$ and $i^t h$ parts $\geq k_i$. Let $D_m^{(l,N)}(v)$ denote the number of compositions of v enumerated by $D^{(l,N)}(v)$ into exactly m parts.

Theorem 3.

$$\text{(i)} \qquad C_m^{(l,N)}(v) = D_m^{(l,N)}(v)$$

and

$$\text{(ii)} \qquad C^{(l,N)}(v) = D^{(l,N)}(v).$$

Proof: First we deduce a generating function for $C_m^{(l,N)}(\nu,n)$:

$$\sum_{\nu,n\geq l} C_m^{(l,N)}(\nu,n)z^\nu q^n = \frac{z^{k_2+k_3+\cdots+k_m+l}q^{k_2+2k_3+\cdots+(m-1)k_m+ml}}{(z^Nq^N;q^N)_m}. \tag{1}$$

Clearly,

$$\sum_{\nu,n\geq l} C_m^{(l,N)}(\nu,n)z^\nu q^n = \sum_{n_1,n_2,\cdots,n_m\geq 0} z^{Nn_1+(m-1)+l}q^{(Nn_m+l)+(Nn_{m-1}+1+l)+\cdots+(Nn_1+(m-1)+l)} \tag{2}$$

where,

$$n_{j-1}-n_j \geq \frac{k_j-1}{N}, \qquad j=2,3,\ldots,m. \tag{3}$$

Setting

$$p_{j-1} := n_{j-1}-n_j-\frac{k_j-1}{N}, \qquad j=2,3,\ldots,m,$$

we have

$$p_1+p_2+p_3+\cdots+p_{m-1} = (n_1-n_m)-\frac{(k_2+\cdots+k_m)}{N}+\frac{m-1}{N}.$$

and

$$\begin{aligned} p_1+2p_2+3p_3+\cdots+(m-1)p_{m-1} &= (n_1+n_2+\cdots+n_m)-mn_m \\ &\quad -\frac{k_2+2k_3+\cdots+(m-1)k_m}{N}+\frac{m(m-1)}{2N}. \end{aligned}$$

Using these in (2) the right side of (2) becomes

$$\begin{aligned} &z^{l+k_2+\cdots+k_m}\cdot q^{k_2+2k_3+\cdots+(m-1)k_m+ml}\cdot\sum_{p_1\geq 0}(zq)^{2p_1}\cdot\sum_{p_2\geq 0}(zq^2)^{2p_2}\cdots \\ &\sum_{p_{m-1}\geq 0}(zq^{m-1})^{2p_{m-1}}\cdot\sum_{n_m\geq 0}(zq^m)^{2n_m} \end{aligned}$$

which reduces to the right side of (1).

From (1) we have, on letting $q\to 1$,

$$\begin{aligned} \sum_{\nu\geq l} C_m^{(l,N)}(\nu)z^\nu &= z^{l-(m-1)(k_2+\cdots+k_m)}\left[\frac{z^{(k_2+\cdots+k_m)}}{1-z^N}\right]^m \\ &= z^{l-(m-1)(k_2+\cdots+k_m)}[z^{k_2+\cdots+k_m}+z^{k_2+\cdots+k_m+N}+\cdots]^m \\ &= \sum_{\nu\geq l} D_m^{(l,N)}(\nu)z^\nu. \end{aligned} \tag{4}$$

Comparing the coefficients of z^ν we obtain part (i) of the theorem. Part (ii) follows immediately from part (i). □

3 Alternative proof by MacMahon's Partition Analysis

Following MacMahon's analysis we can rewrite the right side of (2) along with the conditions (3) as

$$\Omega_{\geq} \sum_{n_1,\ldots,n_m\geq 0} z^{Nn_1+(m-1)+l} q^{(Nn_m+l)+\cdots+(Nn_1+(m-1)+l)} \lambda_1^{n_1-n_2-\frac{k_2-1}{N}} \cdots \lambda_{m-1}^{n_{m-1}-n_m-\frac{k_m-1}{N}}. \tag{5}$$

Here $\Omega_{\geq}$ is a linear operator operating on the Laurent series in $\lambda_1, \lambda_2, \ldots, \lambda_{m-1}$; it annihilates terms with negative exponents and sets each $\lambda_j = 1$ in the remaining terms. Thus under the condition (3), the expression (5) equivalently the right side of (2), becomes

$$z^{(m-1)+l} \cdot q^{\frac{m}{2}(m-1+2l)} \cdot \Omega_{\geq} \sum_{n_1,\ldots,n_m\geq 0} (z^N q^N \lambda_1)^{n_1} \left(\frac{q^N \lambda_2}{\lambda_1}\right)^{n_2} \cdots$$

$$\left(\frac{q^N}{\lambda_{m-1}}\right)^{n_m} \lambda_1^{-\frac{k_2-1}{N}} \cdots \lambda_{m-1}^{-\frac{k_{m-1}-1}{N}} = z^{(m-1)+l} \cdot q^{\frac{m}{2}(m-1+2l)} \cdot \Omega_{\geq}$$

$$\frac{\lambda_1^{-\frac{k_2-1}{N}} \cdots \lambda_{m-1}^{-\frac{k_{m-1}-1}{N}}}{(1-z^N q^N \lambda_1)\left(1-\frac{q^N \lambda_N}{\lambda_1}\right)\cdots\left(1-\frac{q^N \lambda_{m-1}}{\lambda_{m-2}}\right)\left(1-\frac{q^N}{\lambda_{m-1}}\right)}. \tag{6}$$

Applying repeatedly the following standard result :

$$\Omega_{\geq} \frac{\lambda^{-\alpha}}{(1-\lambda x)(1-y/\lambda)} = \frac{x^{\alpha}}{(1-x)(1-xy)}.$$

(6) reduces to the right side of (1) The rest of the proof of Theorem 3 is as given in Section 2.

References

[1] A.K. Agarwal, An Analogue of a Rogers-Ramanujan identity, *4th Int. Conf. SSFA*, Vol.4, 2003, pp. 9-12.

[2] S. Bhargava, Chandrashekar Adiga and N. Anitha, On a class of N-parity Partition, (preprint)

[3] P.A. MacMahon, *Combinatory Analysis*, Vol.2, Cambridge University Press, 1960.

Chandrashekar Adiga, N. Anitha and Jung Hun Han
Department of Studies in Mathematics,
University of Mysore,
Manasa Gangotri,
Mysore 570 006, India.

The Riemann Zeta Function and Related Themes – 2006, pp. 19–27

Probabilistic number theory and random permutations: Functional limit theory

Gutti Jogesh Babu

Dedicated to Professor K. Ramachandra on his 70th birthday

Abstract

The ideas from Probabilistic Number Theory are useful in the study of measures on partitions of integers. Connection between the Ewens sampling formula in population genetics and the partitions of an integer generated by random permutations will be discussed. Functional limit theory for partial sum processes induced by Ewens sampling formula is reviewed. The results on limit processes with dependent increments are illustrated.

1 Introduction

In the last few decades, mathematical population geneticists have been exploring the mechanisms that maintain diversity in a population. In 1972, Ewens established a formula to describe the probability distribution of a sample of genes from a population that has evolved over many generations, by a family of measures on the set of permutations of the first n integers (equivalently on the set of partitions of n). The Ewens formula can be used to test if the popular assumptions are consistent with data, and to estimate the parameters. The statistics that are useful in this connection will generally be expressed as functions of the sums of transforms of the 'allelic partition'. Such statistics can be viewed as functions of a process on the permutation group of integers.

In a series of papers [3]-[7], Babu and Manstavičius, have developed necessary and sufficient conditions for the weak convergence of a partial sum process based on these measures to a process with independent increments. Under very general conditions, it has been shown that a partial sum process converges weakly in a function space if and only if a related process defined through sums of independent random variables converge. In this paper, the case where the limiting processes need not be processes with independent increments is considered. Thus, under Ewens sampling formula, the limiting process of the partial sums of dependent variables differs from that of the associated process defined through the partial sums of independent random variables. The basic ideas for proofs come from probabilistic number theory and analytic number theory. Integration over Hankel contour (see Corollary 2.1 in Section II.5.2 in [17]) plays an important role.

1991 *Mathematics Subject Classification.* 60F17, 60C05, 11K65.

Key words and phrases. Cycle, Ewens sampling formula, functional limit theorem, random partitions, slowly varying function, weak convergence.

Research of Babu was supported in part by NSF grants DMS-0101360, AST-0434234 and AST-0326524.

2 Probabilistic Number Theory

We shall start with a brief comparative analysis of the developments in probabilistic number theory and the theory of random permutations. The uniform probability measure

$$\nu_n(A) = \frac{1}{n}\#\{1 \le m \le n : m \in A\}$$

on integers, satisfies for $k, l \ge 0$,

$$\nu_n(\alpha_p(m) = k) = \frac{1}{n}\left(\left[\frac{n}{p^k}\right] - \left[\frac{n}{p^{k+1}}\right]\right) \approx \frac{1}{p^k}\left(1 - \frac{1}{p}\right), \qquad k \ge 0,$$

and

$$\nu_n(\alpha_p(m) = k, \alpha_q(m) = l) \approx \frac{1}{p^k}\left(1 - \frac{1}{p}\right)\frac{1}{q^l}\left(1 - \frac{1}{q}\right),$$
$$\approx \nu_n(\alpha_p(m) = k)\nu_n(\alpha_q(m) = l),$$

where $m = \prod_p p^{\alpha_p(m)}$ is the unique representation of integer m as the product of prime powers. It follows that α_p has asymptotically geometric distribution. In addition, α_p and α_q are asymptotically independent, where p and q are distinct primes.

The Fundamental Theorem of probabilistic number theory [15] states that

$$\nu_n(\alpha_p(m) = k_p, p \le r) = P(\xi_p = k_p, p \le r) + o(1),$$

where $r \le n^\varepsilon$ for each $\varepsilon > 0$, and $\{\xi_p\}$ are independent geometric random variables

$$P(\xi_p = k) = \frac{1}{p^k}\left(1 - \frac{1}{p}\right), \qquad k \ge 0.$$

If h is an additive arithmetic function, $h(mn) = h(m) + h(n)$, $(m, n) = 1$, then h can be represented as $h(m) = \sum_p h(p^{\alpha_p(m)})$. As $\alpha_p(m) = 1$ for a prime $\sqrt{n} < p \le n$ implies $\alpha_q(m) = 0$ for all primes $q \ne p$, $\sqrt{n} < q \le n$, it follows that they are not independent even asymptotically. To establish the limiting distribution of h under ν_n, one uses the decomposition $h(m) = h_r(m) + h^r(m)$, where

$$h_r(m) = \sum_{p \le r} h(p^{\alpha_p(m)}), \quad h^r(m) = \sum_{p > r} h(p^{\alpha_p(m)}).$$

Then the fundamental theorem of probabilistic number theory is used to approximate $\nu_n(h_r(m) \le x)$ by $P\left(\sum_{p\le r} f_p(\xi_p) \le x\right)$ and showing that the contribution of h^r is negligible, where $f_p(k) = h(p^k)$.

Similar ideas are used in obtaining functional limit theorems by Babu [2] for the partial sum process

$$X_n(t) = \left(1/\sqrt{B(n,n)}\right)\sum{}' h(q), \quad B(n,k) = \sum_{p \le k} \frac{1}{p} h^2(p), \quad t \in [0,1],$$

where the sum $\sum'$ is taken over all primes $q \le n$ satisfying $B(n, q) \le tB(n, n)$.

3 Statistical Group Theory

Similar approach can be used in the study of statistical group theory and in particular random permutations. Let $\mathbf{S}_n$ denote the group of permutations on $\{1, \ldots, n\}$ Each permutation σ can be decomposed as $\sigma = \kappa_1 \ldots \kappa_\omega$ where $\omega(\sigma)$ denotes the number of cycles of σ, and κ_i denote the independent cycles. For example, the permutation τ that maps $\{1, 2, 3, 4, 5, 6, 7, 8\}$ to $\{5, 3, 6, 1, 8, 2, 7, 4\}$ has three cycles (1 5 8 4), (2 3 6), (7) and can be represented as $\tau = (1\ 5\ 8\ 4)\ (2\ 3\ 6)\ (7)$. Thus $Ord(\tau) = 12$, where the order $Ord(\sigma)$ of permutation σ is defined to be the smallest k such that σ^k = identity permutation. If $k_j(\sigma)$ denotes the number of cycles of length j of σ, then $\omega(\sigma) = k_1(\sigma) + \cdots + k_n(\sigma)$ and $Ord(\sigma) = \text{l.c.m.}\{j \le n : k_j(\sigma) > 0\}$.

Goncharov, Erdos-Turan, and others contributed to the theory. In 1942, V. L. Goncharov [12] has shown that

$$\frac{1}{n!}\#\left\{\sigma \in \mathbf{S}_n : \omega(\sigma) - \log n < x\sqrt{\log n}\right\} \to \Phi(x),$$

where $\Phi(x) = \frac{1}{\sqrt{2\pi}} \int_0^x e^{-\frac{1}{2}u^2} du$. In 1965, Erdös and Turán [10] have established that

$$\frac{1}{n!}\#\left\{\sigma \in \mathbf{S}_n : \log Ord(\sigma) - \frac{1}{2}\log^2 n \le \frac{x}{\sqrt{3}}\log^{3/2} n\right\} \to \Phi(x).$$

However, in these and other early works, there is no trace of the use of ideas from probabilistic number theory, though the functions are similar.

The equivalent relation, $\sigma \sim \tau$ if $k_j(\sigma) = k_j(\tau)$ for all j, partitions $\mathbf{S}_n$ into equivalence classes, known as conjugate classes. Hence we can identify σ with the vector $\bar{k} = (k_1(\sigma), \ldots, k_n(\sigma))$, where $1k_1(\sigma) + \cdots + nk_n(\sigma) = n$. This leads to random partitions of integer n.

4 Ewens Sampling Formula

The family of probability measures on the symmetric group $\mathbf{S}_n$ of permutations on $\{1, \ldots, n\}$, induced by the Ewens sampling formula (see [11]) are given by

$$\nu_{n,\theta}(\bar{k}) := \frac{n!}{\theta_{(n)}} \prod_{j=1}^{n} \left(\frac{\theta}{j}\right)^{k_j} \frac{1}{k_j!}, \quad \bar{k} := (k_1, \ldots, k_n) \in \mathbf{Z}^{+^n},$$

for the partition $n = 1k_1 + \cdots + nk_n$, $n \in \mathbf{N}$, and 0 otherwise, where $\theta > 0$, and $\theta_{(n)} = \theta(\theta+1)\cdots(\theta+n-1)$. The quantity $\nu_{n,\theta}(\bar{k})$ can also be viewed as the probability measure on the class of conjugate elements $\sigma \in \mathbf{S}_n$, all having $k_j(\sigma) = k_j$ cycles of length j, $1 \le j \le n$. The probability measure $\nu_{n,\theta}$ is induced by the measure $\nu'_{n,\theta}$ on $\mathbf{S}_n$, that assigns a mass proportional to $\theta^{w(\sigma)}$ for $\sigma \in \mathbf{S}_n$, where $w(\sigma) = k_1(\sigma) + \cdots + k_n(\sigma)$ denotes the total number of cycles of σ. This can be seen from

$$\nu'_\theta(\sigma) = \theta^{w(\sigma)} \left(\sum_{\tau \in \mathbf{S}_n} \theta^{w(\tau)}\right)^{-1} = \frac{\theta^{w(\sigma)}}{\theta_{(n)}}.$$

Thus, we use this probability measure on $\mathbf{S}_n$ and leave the same notation $\nu_{n,\theta}$ for it.

The case $\theta = 1$ corresponds to the measure induced by the uniform probability $(1/n!)\#\{\sigma \in \mathbf{S}_n : \cdots\}$ on $\mathbf{S}_n$. If $k_j(\sigma) = 1$ for some $\frac{n}{2} < j \le n$, then $k_i(\sigma) = 0$ for all $\frac{n}{2} < i \le n, i \ne j$.

As mentioned in the introduction §1, the Ewens formula describes the probability law of a sample of n genes from a population that has evolved over many generations. The domain of $\nu_{n,\theta}$, $\{(k_1, \ldots, k_n) : 1k_1 + \cdots + nk_n = n\}$ is same as that of the Allelic Partition $\underline{k} = (k_1, \ldots, k_n)$, where k_j denotes the number of alleles appearing j times. The distribution of *Allelic Partition* has all the information available in the sample of n genes. Hence, the Ewens formula can be used to test if the popular assumptions are consistent with data, and to estimate the parameters.

This motivates consideration of additive functions on $\underline{\mathbf{S}}_n$. A function $h : \mathbf{S}_n \to \mathbf{R}$ is called additive if $h_j(0) = 0$ and $h(\sigma) = \sum_{j=1}^{n} h_j(k_j(\sigma))$, $\sigma \in \underline{k}$. Kolchin and Chistyakov [14] showed that $\nu_{n,1}(\sum_{j\le r} a_{jn}k_j(\sigma) - A_r < x)$ converges for some sequence $\{A_r\}$ if and only if $P(\sum_{j\le r} a_{jn}Y_j - A_r < x)$ converges, where Y_j are independent Poisson random variable with mean $1/j$, $r = r(n) \to \infty$, and $r \log r = o(n)$.

To facilitate the study of the limiting distributions of additive functions on $\mathbf{S}_n$, Arratia and Tavaré [1] developed a result similar to fundamental result of Kubilius in probabilistic number theory, which states that

$$\nu_{n,1}(k_j(\sigma) = k_j, j \le r) = P(Y_j = k_j, j \le r) + O_\delta\left(exp\left\{-(1-\delta)\frac{n}{r}\log\frac{n}{r}\right\}\right).$$

The measure $\nu_{n,\theta}$ can be represented using independent Poisson random variables ξ_j with $\mathbf{E}(\xi_j) = \frac{\theta}{j}$, as

$$\nu_{n,\theta}(\overline{k}) = P(\xi_1 = k_1, \ldots, \xi_n = k_n \,|\zeta = n),\ \zeta = 1\xi_1 + \cdots + n\xi_n.$$

5 Functional Limit Theorems: Processes with independents

To state the functional limit theorems, let $h(\sigma) = \sum_{i=1}^{n} h_j(k_j(\sigma))$ denote an additive function on $\mathbf{S}_n$. Let

$$A(u) = \theta \sum_{j\le u} \frac{1}{j}\, h_j(1),\ B^2(u) = \theta \sum_{j\le u} \frac{1}{j}\, h_j(1)^2,$$

and

$$y_n(t) = \max\{u : B^2(u) \le tB^2(n)\}.$$

The first functional limit theorem for $\theta = 1$ and $\omega(\sigma)$ was obtained by DeLaurentis and Pittel [8]. This was extended to general θ by Hansen [13] and Donnelly *et al.* [9]. The following theorem for general additive functions is from [3].

Theorem 1 (Babu and Manstavičius [3]). *Suppose* $B(n) \to \infty$, *and*

$$H_n(\sigma, t) = \frac{1}{B(n)}\left(\sum_{j\le y_n(t)} h_j(k_j(\sigma)) - A(y_n(t))\right).$$

Then $\nu_{n,\theta} \cdot H_n^{-1} \Rightarrow W$ *if and only if for each* $\varepsilon > 0$,

$$\Lambda_n(\varepsilon) = \frac{1}{B^2(n)} \sum_{|h_j(1)|\ge \varepsilon B(n)} \frac{1}{j}\, h_j(1)^2 \to 0,$$

where W *denotes the Brownian Motion on* $[0, 1]$.

This result leads, via invariance principle of the probability theory, to the limiting distributions of functions of partial sums of h_j These results throw new light on the partitions of integers. For example the functional limit theorem implies,

$$\nu_{n,\theta}(h(\sigma) - A(n) \leq xB(n)) \to \Phi(x),$$

$$\nu_{n,\theta}\left(\sup_{k\leq n}\left(\sum_{j\leq k} h_j(k_j(\sigma)) - A(k)\right) \leq xB(n)\right) \to P\left(\sup_{0\leq t\leq 1} W(t) \leq x\right)$$
$$= 2\Phi(x), \qquad x \geq 0$$

$$\nu_{n,\theta}\left(\sup_{k\leq n}\left|\sum_{j\leq k} h_j(k_j(\sigma)) - A(k)\right| \leq xB(n)\right) \to P\left(\sup_{0\leq t\leq 1} |W(t)| \leq x\right)$$
$$= \frac{4}{\pi}\sum_{k=0}^{\infty}\frac{(-1)^k}{2k+1}e^{-\pi^2(2k+1)^2/8x^2}.$$

By using a slowly varying function to scale h, Babu and Manstavičius obtained convergence to a stable processes and to general processes with independent increments. Let $\beta(n) \to \infty$ and $x^* = \min(|x|, 1)\,\mathrm{sign}(x)$. For an additive function h, let

$$c(u, n) = \sum_{j\leq u}\frac{1}{j}\left(\frac{h_j(1)}{\beta(n)}\right)^{*2}, \quad A(u, n, \beta) = \theta\sum_{j\leq u}\frac{1}{j}\left(\frac{h_j(1)}{\beta(n)}\right)^{*},$$

$s_n(t) = \max\{l \leq n : c(l, n) \leq t\, c(n, n)\}$,

$$R_{n,h}(t) = \frac{1}{\beta(n)}\sum_{j\leq s_n(t)} h_j(k_j(\sigma)) - A(s_n(t), n, \beta)$$

and

$$X_{n,h}(t) = \frac{1}{\beta(n)}\sum_{j\leq s_n(t)} h_j(1)\xi_j - A(s_n(t), n, \beta).$$

The following result is from Babu and and Manstavičius [7].

Theorem 2. *In order that $R_{n,h} \Rightarrow X$, where X is a process with independent increments and the distribution of $X(1)$ is non-degenerate, it is necessary and sufficient that $X_{n,h} \Rightarrow X$ and $\beta(n)$ is slowly varying.*

If $X_{n,h} \Rightarrow X$, then the limiting process X is necessarily a process with independent increments and it satisfies $P(X(0) = 0) = 1$. It is interesting to note that the convergence of the process defined through the partial sums of dependent random variables is equivalent to the convergence of the process defined through the partial sums of the corresponding independent random variables. The result holds in spite of the strong dependent structure on $\{k_j(\sigma) : \frac{1}{2}n \leq j \leq n\}$.

Counter Example. In the one-dimensional case, $X_{n,h}(1)$ and $R_{n,h}(1)$ need not have the same limit. To see this, let $\theta = 1$, $0 < \alpha < 2$ and let F be the stable law with characteristic function $\phi_\alpha(s) = e^{-|s|^\alpha}$. Let γ_j denote the fractional part of $j\sqrt{2}$,

$$a(j) = \begin{cases} j^{1/\alpha}F^{-1}(\gamma_j) & \text{if } |F^{-1}(\gamma_j)| \leq j^{1/\alpha} \\ 0 & \text{otherwise,} \end{cases}$$

$h(\sigma) = \sum_{j=1}^{n} k_j(\sigma)a(j)$, and $\beta(n) = n^{1/\alpha}$. Then $(h(\cdot)/\beta(n)) - A(n,n,\beta) \Rightarrow F$. But $X_{n,h}(1) \Rightarrow G$, where the characteristic function of G is

$$\phi(s) = \exp\left\{\int_0^1 \frac{1}{y}\left(e^{-y|s|^\alpha} - 1\right)dy\right\}.$$

6 Limit processes with dependent increments

The example above illustrates that if β is not slowly varying function, the limit process may have dependent increments. To study such limits, a generalization of the Main Lemma in [16] that involves integration over Hankel contour of the type given in Figure 1 is needed.

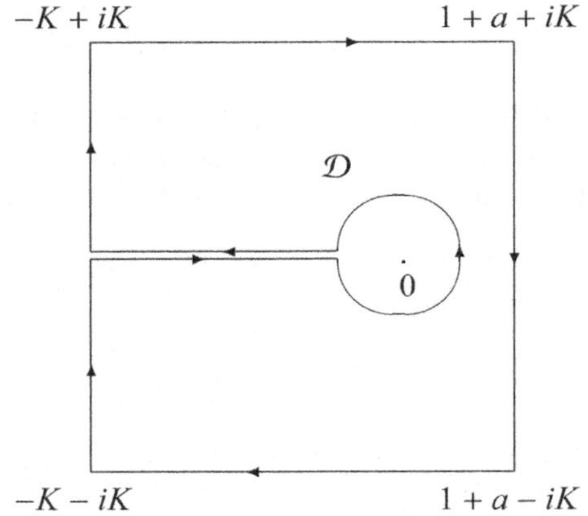

Figure 1: Contour with $a \geq 0$ and $K > 0$

However, to facilitate the discussion we consider an example with $\beta(n) = n^\rho, \rho > 0$. Let the additive function h on $\mathbf{S}_n$ be given by $h(\sigma) = \sum_{j=1}^{n} k_j(\sigma)j^\rho$. Let the processes H_n based on h be given by,

$$H_n(t) := H_n(t,\sigma) = (1/\beta(n)) \sum_{j\leq nt} k_j(\sigma)j^\rho = \sum_{j\leq nt} k_j(\sigma)(j/n)^\rho, \qquad 0 \leq t \leq 1.$$

Note that $H_n(1,\sigma) = 1$ for all $\sigma \in \mathbf{S}_n$ if $\rho = 1$. We now present preliminary notation and results needed in illustrating the limiting process. We restrict to the case $\theta = 1$. First, we shall consider the mean values of multiplicative functions $g : \mathbf{S}_n \to \mathbf{C}$ defined via $g(\sigma) = \prod_{j=1}^{n} f(j)^{k_j(\sigma)}$, where $f(j)$, $j \geq 1$ are complex numbers that may depend on n or other parameters. Its mean value with respect to the measure $\nu_{n,1}$ equals

$$M_n(g) := \frac{1}{n!}\sum_{\sigma\in\mathbf{S}_n} g(\sigma) = \sum_{\substack{k_1,\ldots,k_n\geq 0 \\ 1k_1+\cdots+nk_n=n}} \prod_{j=1}^{n}\left(\frac{1}{j}f(j)\right)^{k_j}\frac{1}{k_j!}.$$

The behavior of $M_n(g)$ is examined in the next Lemma (see [3]) for large cycles.

Lemma 1. *Let $g : \mathbf{S}_n \to \mathbf{C}$ be a multiplicative function defined via f such that $f(j) = 1$ for all but $j \in J \subset (n/2, n]$. Then*

$$M_n(g) = 1 + \sum_{j \in J} \frac{1}{j}(f(j) - 1).$$

Proof: Observe that, if $k_j \geq 1$ for some $j \in J$, then $1k_1 + \cdots + nk_n = n$ implies $k_j = 1$ and $k_l = 0$ for the remaining $l \neq j$ and $l \in J$. Let $\Sigma_{(0)}$ denotes the sum over all $(k_1, \ldots, k_n)$ satisfying $1k_1 + \cdots + nk_n = n$ and $k_l = 0$ for all $l \in J$, and $\Sigma_{(j)}$ denotes the sum over all $(k_1, \ldots, k_n)$ satisfying $1k_1 + \cdots + nk_n = n$ and $k_j = 1$. Hence

$$\begin{aligned} M_n(g) &= \sum\nolimits_{(0)} \prod_{l=1}^{n} \left(\frac{1}{l}\right)^{k_l} \frac{1}{k_l!} + \sum_{j \in J} f(j) \sum\nolimits_{(j)} \prod_{l=1}^{n} \left(\frac{1}{l}\right)^{k_l} \frac{1}{k_l!} \\ &= 1 + \sum_{j \in J} (f(j) - 1) \sum\nolimits_{(j)} \prod_{l=1}^{n} \left(\frac{1}{l}\right)^{k_l} \frac{1}{k_l!}. \end{aligned}$$

The Lemma follows now as the last sum is $(1/j)$.

The characteristic function $\phi_{n,s,t}$ of $H_n(t) - H_n(s)$ for $\frac{1}{2} < s < t \leq 1$ is given by

$$\phi_{n,s,t}(\eta) = M_n\left(e^{i\eta(H_n(t) - H_n(s))}\right), \qquad \eta \in \mathbf{R}.$$

We apply Lemma 1 with $f(j) = \exp(i\eta(j/n)^\rho)$, $\eta \in \mathbf{R}$, to get

$$\begin{aligned} \phi_{n,s,t}(\eta) &= 1 + \sum_{s < (j/n) \leq t} \frac{1}{j}(e^{i\eta(j/n)^\rho} - 1) \\ &\to 1 + \int_s^t \frac{1}{v}\left(e^{i\eta v^\rho} - 1\right) dv \\ &= 1 + \frac{1}{\rho} \int_{s^\rho}^{t^\rho} \frac{1}{u}\left(e^{i\eta u} - 1\right) du =: \phi_{s,t}(\eta). \end{aligned}$$

If the limiting process of H_n has independent increments, then for all $\frac{1}{2} < s < t < 1$, and $\eta \in \mathbf{R}$,

$$\phi_{s,1}(\eta) = \phi_{s,t}(\eta)\, \phi_{t,1}(\eta). \tag{1}$$

Hence for (1) to hold we must have,

$$\left(\int_{s^\rho}^{t^\rho} \frac{1}{u}\left(e^{i\eta u} - 1\right) du\right) \left(\int_{t^\rho}^{1} \frac{1}{u}\left(e^{i\eta u} - 1\right) du\right) = 0$$

for all $\frac{1}{2} < s < t < 1$ and $\eta \in \mathbf{R}$, which is impossible. This shows that the limiting process of H_n is not a process with independent increments.

The weak convergence of processes for general h and non-slowly varying β will be addressed elsewhere. □

References

[1] R. Arratia and S. Tavaré, The cycle structure of random permutations, *Annals of Probability*, **20** (1992) 1567–1591.

[2] G.J. Babu, A note on the invariance principle for additive functions, *Sankhyā, A* **35** (1973), 3, 307–310.

[3] G.J. Babu and E. Manstavičius, Brownian motion and random permutations, *Sankhyā, A* **61** (1999), 3, 312–327.

[4] G.J. Babu and E. Manstavičius, Random permutations and the Ewens sampling formula in genetics, In: *Probab. Theory and Math. Stat.*, B.Grigelionis *et al.* (Eds), VSP/TEV, Vilnius/Utrecht, 1999, 33–42.

[5] G.J. Babu and E. Manstavičius, Limit theorems for random permutations, In: *Paul Erdös and His Mathematics*, János Bolyai Mathematical Society, 1999, 19–22.

[6] G.J. Babu and E. Manstavičius, Infinitely divisible limit processes for the Ewens sampling formula, *Lithuanian Math. J.* **42** (2002), 3, 232–242.

[7] G.J. Babu and E. Manstavičius, Limit Processes with independent increments for the Ewens sampling formula, *Ann. Inst. Stat. Math.* **54** (2002), 3, 607–620.

[8] J.M. DeLaurentis and B.G. Pittel, Random permutations and the Brownian motion, *Pacific J. Math.* **119** (1985), 287–301.

[9] P. Donnelly, T.G. Kurtz and S. Tavaré, On the functional central limit theorem for the Ewens Sampling Formula, *Ann. Appl. Probab.* **1** (1991), 539–545.

[10] P. Erdös and P. Turán. On some problems of a statistical group theory I. *Z. Wahrsch. Verw. Gebiete* **4** (1965) 175–186.

[11] W.J. Ewens, The sampling theory of selectively neutral alleles, *Theor. Pop. Biol.* **3** (1972), 87–112.

[12] V.L. Goncharov, On the distribution of cycles in permutations. *Dokl. Acad. Nauk SSSR* **35** (1942) 299–301 (Russian).

[13] J.C. Hansen, A functional central limit theorem for the Ewens Sampling Formula, *J. Appl. Probab.* **27** (1990), 28–43.

[14] V.F. Kolchin and V.P. Chistyakov, On the cycle structure of random permutations. *Matem. Zametki*, **18** (1975) 929–938 (Russian).

[15] J. Kubilius, *Probabilistic Methods in the Theory of Numbers*. Translations of Mathematical Monographs, **11**, AMS, Providence, Rhode Island, 1964.

[16] E. Manstavičius, Additive and multiplicative functions on random permutations, *Lith. Math. J.* **36** (1996), 400–408.

[17] G. Tenenbaum, *Introduction to analytic and probabilistic number theory*, Cambridge studies in advanced mathematics **46**, Cambridge University Press, Cambridge, 1995.

Gutti Jogesh Babu
Department of Statistics,
319 Thomas Building,
The Pennsylvania State University,
University Park, PA 16802, USA

email: babu@stat.psu.edu
http://www.stat.psu.edu/ babu
http://astrostatistics.psu.edu

The Riemann Zeta Function and Related Themes – 2006, pp. 29–38

Contributions to the theory of the Lerch zeta-function

R. Balasubramanian, S. Kanemitsu and H. Tsukada

Dedicated to Professor K. Ramachandra on his 70th birthday

In this paper we shall give some improvements over the results contained in the recent book of A. Laurinchikas and Garunkštis [10] on the Lerch zeta-function. Specifically, in §1, we shall give a more fundamental version (and proof) of the functional equation for the Lerch zeta-function, stated as Theorem 3.2 in [10] and in §2 we are going to give a more fundamental proof of the mean square of the Lerch zeta-function $\phi(\xi, x, s)$ with respect to the second variable x over (0,1], on the basis of our new method of using the difference equation [6], which is feasible for generalization to cover possibly the work of Egami and Matsumoto [3].

We shall adopt the notation from [10] (cf. [4]), which we state here for convenience.

$$\Phi(z, a, s) = \sum_{n=0}^{\infty} \frac{z^n}{(n+a)^s}, \qquad |z| < 1$$

— the Hurwitz-Lerch zeta-function.

$$\phi(\xi, a, s) = \Phi(e^{2\pi i\xi}, a, s) = \sum_{n=0}^{\infty} \frac{e^{2\pi i\xi n}}{(n+a)^s}$$

— the Lerch zeta-function or the Lipschitz-Lerch transcendent,

$$\zeta(s, a) = \phi(m, a, s) = \sum_{n=0}^{\infty} \frac{1}{(n+a)^s}, \qquad \sigma > 1, \quad (m \in \mathbb{Z})$$

— the Hurwitz zeta-function,

$$\ell_s(m) = e^{2\pi i\xi}\phi(\xi, 1, s) = \sum_{n=1}^{\infty} \frac{e^{2\pi i\xi n}}{n^s}$$

— the polylogarithm (function),

$$\zeta(s) = \zeta(s, 1) = \ell_s(m) = \phi(m, 1, s) = \sum_{n=1}^{\infty} \frac{1}{n^s} \qquad \sigma > 1, \quad (m \in \mathbb{Z})$$

— the Riemann zeta-function.

We use the following standard formulas, which need no further explanations:

$$e^{2\pi i\xi}\phi(\xi, a+1, s) = \phi_1(\xi, a, s) = \sum_{n=1}^{\infty} \frac{e^{2\pi i\xi n}}{(n+a)^s} = \phi(\xi, a, s) - a^{-s},$$

$$e^{2\pi i\xi}\phi(\xi, 2, s) = e^{-2\pi i\xi}\ell_s(\xi) - 1,$$

$$\frac{\partial}{\partial a}\phi(\xi, a, s) = -s\phi(\xi, a, s+1).$$

We also use the following well-known special functions:

$$\Gamma(s, z) = \int_z^{\infty} e^{-t} t^{s-1}\,dt$$

— the incomplete gamma function of the second kind and

$$\delta(z)$$

— the Dirac delta function.

1 The functional equation

In this section, we shall give fundamental proofs of the functional equation for the Hurwitz zeta-function (stated as Theorem 3.1 [10]) and for the Lerch zeta-function (stated as Theorem 3.2 [10]).

Our result (proof) is novel and fundamental in that it gives the "*niryana*" which is perceived in two different ways, one as the Hurwitz zeta-function and the other as the polylogarithm, and it is in conformity with the situation of the Kubert space [Mil], which is spanned by the Hurwitz zeta-function values for $\sigma > 1$ and by the polylogarithm values for $\sigma < 1$.

The use of the Dirac delta function was started in [7] whose Fourier expansion is more or less equivalent to, but in a sense more fundamental than, the Poisson summation formula.

Theorem 1 (The Hurwitz relation or the functional equation). *Let* $0 < a < 1$. *Then the meromorphic function*

$$F(s) = F(s, a) = \sum_{n=1}^{\infty}\left\{\frac{e^{-2\pi i a n}}{(-2\pi i n)^{1-s}}\Gamma(1-s, -2\pi i a n) + \frac{e^{2\pi i a n}}{(2\pi i n)^{1-s}}\Gamma(1-s, 2\pi i a n)\right\} \tag{1.1}$$

admits two different representations according as $\sigma > 1$ *or* $\sigma < 1$.

If $\sigma > 1$, *then*

$$F(s) = \zeta(s, a) - \frac{1}{2a^s} - \frac{1}{s-1}\frac{1}{a^{s-1}}, \tag{1.2}$$

and for $\sigma < 1$,

$$F(s) = \frac{\Gamma(1-s)}{(2\pi)^{1-s}}\left(e^{\frac{\pi i}{2}(1-s)}\ell_{1-s}(1-a) + e^{-\frac{\pi i}{2}(1-s)}\ell_{1-s}(a)\right) \tag{1.3}$$

$$-\frac{1}{2a^s} - \frac{1}{s-1}\frac{1}{a^{s-1}},$$

implying, in particular, the functional equation

$$\zeta(s, a) = \frac{\Gamma(1-s)}{(2\pi)^{1-s}}\left(e^{\frac{\pi i}{2}(1-s)}\ell_{1-s}(1-a) + e^{-\frac{\pi i}{2}(1-s)}\ell_{1-s}(a)\right). \tag{1.4}$$

Proof: For $\sigma > 1$, we substitute the definition of the incomplete gamma function and make the change of variable $z = at - a$ to obtain

$$F(s) = \sum_{n=1}^{\infty}\left(\int_0^{\infty} e^{2\pi izn}(z+a)^{-s}dz + \int_0^{\infty} e^{-2\pi izn}(z+a)^{-s}dz\right) \tag{1.5}$$

$$= \sum_{n\in\mathbb{Z}}\int_0^{\infty} \delta(z-n)\frac{1}{(z+a)^s}dz - \int_0^{\infty}(z+a)^{-s}\,dz$$

$$= \frac{1}{2a^s}\sum_{n=1}^{\infty}\frac{1}{(n+a)^{-s}} - \frac{1}{s-1}a^{1-s}$$

$$= \sum_{n=0}^{\infty}\frac{1}{(n+a)^s} - \frac{1}{2a^s} - \frac{1}{s-1}\frac{1}{a^{s-1}},$$

which is (1.2).

On the other hand, if $\sigma < 1$, we apply the same argument that we applied in [7], i.e. we complete the incomplete gamma function with the remainder expressed in integral form, to obtain

$$F(s) = \Gamma(1-s)\sum_{n=1}^{\infty}\left(\frac{e^{-2\pi ian}}{(-2\pi in)^{1-s}} + \frac{e^{2\pi ian}}{(2\pi in)^{1-s}}\right) \tag{1.6}$$

$$-\sum_{n=1}^{\infty}\left(e^{-2\pi ian}a^{1-s}\int_0^1 e^{2\pi iatn}t^{-s}\,dt + e^{2\pi ian}a^{1-s}\int_0^1 e^{-2\pi iatn}t^{-s}\,dt\right).$$

By the change of variable $z = at - a$, the second term on the right of (1.6) becomes

$$-\sum_{n=1}^{\infty}\left(\int_{-a}^{0} e^{2\pi izn}(z+a)^{-s}\,dz + \int_{-a}^{0} e^{-2\pi izn}(z+a)^{-s}\,dz\right) \tag{1.7}$$

$$= -\sum_{n\in\mathbb{Z}}\int_{-a}^{0}\delta(z-n)\frac{1}{(z+a)^s}\,dz - \int_{-a}^{0}(z+a)^{-s}\,dz$$

$$= -\frac{1}{2a^s} - \frac{1}{s-1}\frac{1}{a^{s-1}},$$

while the first term is

$$\frac{\Gamma(1-s)}{(2\pi)^{1-s}}\left(e^{\frac{\pi i}{2}(1-s)}\ell_{1-s}(1-a) + e^{-\frac{\pi i}{2}(1-s)}\ell_{1-s}(a)\right). \tag{1.8}$$

Substituting (1.7) and (1.8) in (1.6), we complete the proof.

Theorem 2. *Let $0 < a < 1$ and $0 < w < 1$. Then the holomorphic function*

$$G(s) = G(\xi, a, s) = \sum_{n\in\mathbb{Z}}\frac{e^{-2\pi ia(n+\xi)}}{(-2\pi i(n+\xi))^{1-s}}\Gamma(1-s, -2\pi ia(n+\xi)) \tag{1.9}$$

admits two different representations according as $\sigma > 1$ or $\sigma < 1$. If $\sigma > 1$, then

$$G(s) = \phi(\xi, a, s) - \frac{1}{2a^s}, \tag{1.10}$$

while if $\sigma < 1$, then

$$G(s) = \frac{\Gamma(1-s)}{(2\pi)^{1-s}} \left\{ e^{\frac{\pi i}{2}(1-s)} e^{-2\pi i \xi a} \phi(-a, 1-s, \xi) + e^{-\frac{\pi i}{2}(1-s)} e^{2\pi i(1-\xi)a} \phi(a, 1-\xi, 1-s) \right\} - \frac{1}{2a^s}, \tag{1.11}$$

implying, in particular, the functional equation

$$\phi(\xi, a, s) = \frac{\Gamma(1-s)}{(2\pi)^{1-s}} \left\{ e^{\frac{\pi i}{2}(1-s)} e^{-2\pi i \xi a} \phi(-a, \xi, 1-s) + e^{-\frac{\pi i}{2}(1-s)} e^{2\pi i(1-\xi)a} \phi(a, 1-\xi, 1-s) \right\}. \tag{1.12}$$

Once stated in the form of (1.9), the proof goes on similar lines as those of the proof of Theorem 1.

Remark 1.1. For various proofs of Theorems 1 and 2, we refer to [7], [10] and [15]; [7] and [10], combined together, give a complete list of references. [13] is missing in [7] and [5] and [11] are missing in [10].

2 Structural elucidation of the mean square of the Lerch zeta-function.

The best known result for the mean square of the Lerch zeta-function is Theorem 1 of Katsurada [9], which follows from his Theorem 3 [9], and thus it suffices to consider the latter theorem.

For complex variables u, v, let

$$E^* = \{(u,v) | u+v \in \mathbb{Z}, u+v \le 2\} \cup \{(u,v) | u \in \mathbb{Z}, \ or\ v \in \mathbb{Z}\} \tag{2.1}$$

denote the exceptional set. For any integer $N \ge 1$, let

$$S_N(u,v;\xi) = \sum_{n=0}^{N-1} \frac{(u)_n}{(1-v)_{n+1}} \left\{ e^{-2\pi i \xi} \ell_{u+n}(\xi) - 1 \right\} \tag{1.11$'$}$$

and

$$T_N(u,v;\xi) = \frac{(u)_N}{(1-v)_N} \sum_{\ell=1}^{\infty} \frac{e^{2\pi i \xi \ell}}{\ell^{u+v-1}} \int_1^{\infty} \beta^{u+v-2}(1+\beta)^{-u-v} d\beta, \tag{1.13$'$}$$

where, here and in what follows, we use a special label for the formulas, i.e. one with prime indicates a formula which has its counterpart in [6], and where $(u)_n = u(u+1)\cdots(u+n-1)$.

Theorem 3 (Katsurada [Ka 2, Theorem 3]). *In the region* $-N+1 < \operatorname{Re} u$, $\operatorname{Re} v < N+1$, *except for the points in* E^*, *we have*

$$\int_1^2 \phi(\xi,x,u)\phi(-\xi,x,v)dx = \frac{1}{u+v-1} + \Gamma(u+v-1) \left\{ \ell_{u+v-1}(\xi)\frac{\Gamma(1-v)}{\Gamma(u)} + \ell_{u+v-1}(-\xi)\frac{\Gamma(1-u)}{\Gamma(v)} \right\} - S_N(u,v;\xi) - S_N(v,u;-\xi) - T_N(u,v;\xi) - T_N(u,v;-\xi).$$

We give a structural proof of Theorem 3 on the same lines as those of our previous paper [6]. The main ingredient in the proof is the following relation, which is feasible for generalization ($\triangle$ means the difference operator $\triangle f(x) = f(x+1) - f(x)$)

$$-\triangle\phi(\xi, u, x)\phi(-\xi, v, x) = x^{-v}e^{2\pi i\xi}\phi(\xi, x+1, u) + x^{-u}e^{-2\pi i\xi}\phi(-\xi, x+1, v) + x^{-u-v}. \tag{2.2}$$

With (2.2) in mind, we put for $\kappa \geq 2N$,

$$f_\kappa(\xi, x, u, v) = \phi(\xi, x, u)\phi(-\xi, x, v) - \phi(0, x, u+v) \tag{2.7$'$}$$

$$- \sum_{k=0}^{\kappa} \frac{\Gamma(u+k-1)}{\Gamma(u)k!} e^{2\pi i\xi}\beta_k(\xi)\phi(\xi, x, u+v+k-1)$$

$$- \sum_{k=0}^{\kappa} \frac{\Gamma(v+k-1)}{\Gamma(v)k!} e^{-2\pi i\xi}\beta_k(-\xi)\phi(-\xi, x, u+v+k-1),$$

where

$$\beta_k(\xi) = (-)^k B_k(1, e^{2\pi i\xi}) \tag{2.3}$$

and where $B_k(z, e^{2\pi i\xi})$ signifies the Frobenius-Vandiver-Apostol generalized Bernoulli polynomial ([1], [8] and [14]) defined by

$$\frac{te^{-zt}}{1 - e^{2\pi i\xi}e^{-t}} \left(= \frac{-te^{-zt}}{e^{2\pi i\xi}e^{-t} - 1} \right) \tag{2.4}$$

$$= \sum_{k=0}^{\infty} B_k(z, e^{2\pi i\xi}) \frac{(-t)^k}{k!}, \qquad |-t + 2\pi i\xi| < 2\pi.$$

Then, corresponding to (2.8) and (2.9) of [6],

$$-\triangle f_\kappa(\xi, u, v, x) = g_\kappa(\xi, u, v, x) + g_\kappa(-\xi, u, v, x), \tag{2.8$'$}$$

where

$$g_\kappa(\xi, u, v, x) = e^{2\pi i\xi}x^{-v}\left(\phi(\xi, x+1, u) - \sum_{k=0}^{\kappa} \frac{\Gamma(u+k-1)}{\Gamma(u)k!}\beta_k(\xi)x^{-u-k+1}\right), \tag{2.9$'$}$$

$$g_\kappa(-\xi, u, v, x) = e^{-2\pi i\xi}x^{-u}\left(\phi(-\xi, x+1, v) - \sum_{k=0}^{\kappa} \frac{\Gamma(v+k-1)}{\Gamma(v)k!}\beta_k(-\xi)x^{-v-k+1}\right).$$

We now quote the following results of Katsurada.

Lemma 1 ([9], (4.4)). *For any integer $K \geq 1$,*

$$T_N(u, v; \xi) = \sum_{k=1}^{K} (-1)^{k-1} \frac{(2-u-v)_{k-1}(u)_{N-k}}{(1-v)_N} \sum_{\ell=1}^{\infty} \frac{e^{2\pi i\xi\ell}}{\ell^k(\ell+1)^{u+N-k}}$$

$$+ (-1)^K \frac{(2-u-v)_K(u)_{N-k}}{(1-v)_N} \sum_{\ell=1}^{\infty} \frac{e^{2\pi i\xi\ell}}{\ell^{u+v-1}}$$

$$\int_\ell^\infty \beta^{u+v-K-2}(1+\beta)^{u+N-k}d\beta.$$

Lemma 2 ([Ka 1], (2.2)). *For any integer $K \geq 1$, and the complex variable z in the region $|\arg z| < \pi$ and $\sigma > -K$, we have*

$$\phi(\xi, \alpha + z, s) = \sum_{k=0}^{K-1} \frac{(s)_k}{(k+1)!} \beta_{k+1}(\xi) z^{-s-k} + O\left(|z|^{-\mathrm{Re}\, z - K}\right).$$

Lemma 2 assures that

$$g_\kappa(\xi, u, v, x) = O(x^{-\mathrm{Re}\, u - \mathrm{Re}\, v - \kappa})$$

and *a fortiori* that

$$\triangle f_\kappa(\xi, u, v, x) = O(x^{-\mathrm{Re}\, u - \mathrm{Re}\, v - \kappa}),$$

enabling us to telescope the series.

We further need a counterpart of Lemma 3 [6]. Recall the representation [4, p.27, (3)] or [10, (2.6), p.20] ($\sigma > 0, \xi \notin \mathbb{Z}$)

$$\begin{aligned} \phi(\xi, \alpha + z, s) &= \Phi(e^{2\pi i \xi}, \alpha + z, s) \\ &= \frac{1}{\Gamma(s)} \int_0^\infty \frac{t^{s-1} e^{-(\alpha+z)t}}{1 - e^{2\pi i \xi} e^{-t}} \, dt \\ &= \frac{1}{\Gamma(s)} \int_0^\infty \frac{t e^{-zt}}{1 - e^{2\pi i \xi} e^{-t}} e^{-\alpha t} t^{s-2} \, dt. \end{aligned}$$

We divide the interval $(0, \infty)$ into two parts: $(0, 1)$ and $[1, \infty)$ and subtract the partial sum of the generating function (2.4) for $B_k(z, e^{2\pi i \xi})$ to obtain an analogue of Lemma 3 [6] whose special case is stated as Lemma 3:

Lemma 3. *Suppose $\sigma > -M, s \neq 1, \xi \notin \mathbb{Z}$ and $x > -1$, where $M \geq 1$ is an integer. Then*

$$\begin{aligned} \phi(\xi, x+1, s) = {} & \frac{1}{\Gamma(s)} \int_1^\infty \frac{e^{-xt}}{e^t - e^{2\pi i \xi}} t^{s-1} \, dt \qquad (3.5') \\ & + \frac{1}{\Gamma(s)} \int_0^1 \left(\frac{t}{e^t - e^{2\pi i \xi}} - \sum_{k=0}^{M} \frac{\beta_k(\xi)}{k!} t^k \right) e^{-xt} t^{s-2} \, dt \\ & + \sum_{k=0}^{M} \frac{\Gamma(s+k-1)}{\Gamma(s) k!} \beta_k(\xi) x^{-s-k+1} - \sum_{k=0}^{M} \frac{\Gamma(s+k-1, x)}{\Gamma(s) k!} \beta_k(\xi) x^{-s-k+1}. \end{aligned}$$

We have also ($\xi \notin \mathbb{Z}, \sigma > -M, s \neq 1$)

$$\begin{aligned} \phi(\xi, 1, s) = {} & \frac{1}{\Gamma(s)} \int_1^\infty \frac{t}{e^t - e^{2\pi i \xi}} t^{s-1} \, dt \qquad (3.3)' \\ & + \frac{1}{\Gamma(s)} \int_0^1 \left(\frac{t}{e^t - e^{2\pi i \xi}} - \sum_{k=0}^{M} \frac{\beta_k(\xi)}{k!} t^k \right) t^{s-2} \, dt \\ & + \sum_{k=0}^{M} \frac{\beta_k(\xi)}{\Gamma(s) k!} \frac{1}{s+k-1}. \end{aligned}$$

We now turn to the proof of Theorem 3. By integrating (2.9)′ N-times, we obtain

$$\int_1^\infty g_\kappa(\xi, u, v, x) \, dx = -S_N(u, v; \xi) + S_2(\xi, u) + S_3(\xi, u), \qquad (2.10)'$$

where

$$S_2(\xi, u) = \sum_{n=0}^{N-1} \frac{e^{2\pi i\xi}}{(1-v)_{n+1}} \sum_{k=0}^{\kappa} \frac{\Gamma(u+n+k-1)}{\Gamma(u)k!}\beta_k(\xi) \tag{2.5}$$

and

$$S_3(\xi, u) = \frac{(u)_N}{(1-v)_N} \int_1^\infty g_\kappa(\xi, u+N, v-N, x)\, dx. \tag{2.6}$$

We may apply the same argument that we applied in [6] to $S_2(\xi, u)$ and $S_3(\xi, u)$, to obtain

$$S_2(\xi, u) = \frac{1}{(1-v)_N} \sum_{k=0}^{\kappa} \frac{\Gamma(u+N+k-1)}{\Gamma(u)k!} \frac{e^{2\pi i\xi}\beta_k(\xi)}{u+v+k-2} \tag{3.8$'$}$$

$$- \sum_{k=0}^{\kappa} \frac{\Gamma(u+k-1)}{\Gamma(u)k!} \frac{e^{2\pi i\xi}\beta_k(\xi)}{u+v+k-2},$$

and

$$S_3(\xi, u) = \frac{e^{2\pi i\xi}(u)_N}{(1-v)_N} \left[\frac{1}{\Gamma(u+N)} \int_1^\infty dx \int_1^\infty \frac{e^{-xt}}{e^t - e^{2\pi i\xi}} t^{u+N-1} x^{N-v}\, dt \right. \tag{3.9$'$}$$

$$+ \frac{1}{\Gamma(u+N)} \int_1^\infty dx \int_0^1 \left(\frac{t}{e^t - e^{2\pi i\xi}} - \sum_{k=0}^{\kappa} \frac{\beta_k(\xi)}{k!} t^k \right) e^{-xt} t^{u+N-1} x^{N-v} dt$$

$$\left. - \sum_{k=0}^{\kappa} \frac{\beta_k(\xi)}{\Gamma(u+N)k!} \int_1^\infty x^{-u-v-k+1} \Gamma(u+N+k-1, x)\, dx \right]$$

$$= \frac{e^{2\pi i\xi}(u)_N}{(1-v)_N} (S_{3,1} + S_{3,2} + S_{3,3}),$$

say. The same method of completing the integral over $(1, \infty)$ in x to one over $(0, \infty)$ applies and we have

$$S_{3,1} = \frac{\Gamma(1-v+N)}{\Gamma(u+N)} \int_1^\infty \frac{t^{u+v-2}}{e^t - e^{2\pi i\xi}}\, dt \tag{3.10$'$}$$

$$- \frac{1}{\Gamma(u+N)} \int_0^1 \int_1^\infty \frac{e^{-xt}}{e^t - e^{2\pi i\xi}} t^{u+N-1} x^{N-v}\, dt\, dx,$$

$$S_{3,2} = \frac{\Gamma(1-v+N)}{\Gamma(u+N)} \int_0^1 \left(\frac{t}{e^t - e^{2\pi i\xi}} - \sum_{k=0}^{\kappa} \frac{\beta_k(\xi)}{k!} t^k \right) t^{u+v-2}\, dt \tag{3.11$'$}$$

$$- \frac{1}{\Gamma(u+N)} \int_0^1 \int_0^1 \left(\frac{t}{e^t - e^{2\pi i\xi}} - \sum_{k=0}^{\kappa} \frac{\beta_k(\xi)}{k!} t^k \right) e^{-xt} t^{u+N-2}\, dt\, dx,$$

and

$$S_{3,3} = \frac{\Gamma(1-v+N)}{\Gamma(u+N)} \sum_{k=0}^{\kappa} \frac{\beta_k(\xi)}{k!} \frac{1}{u+v+k-2} \tag{3.12$'$}$$

$$- \frac{\Gamma(u+N+k-1)}{\Gamma(u+N)} \sum_{k=0}^{\kappa} \frac{\beta_k(\xi)}{k!} \frac{1}{u+v+k-2}$$

$$+ \frac{1}{\Gamma(u+N)} \int_0^1 \sum_{k=0}^{\kappa} (\Gamma(u+N+k-1) - \Gamma(u+N+k-1, x)) \frac{\beta_k(\xi)}{k!} x^{-u-v-k+1}\, dx.$$

Substituting (3.10)′ - (3.12)′ in ′, we obtain

$$S_3 = \frac{(u)_N}{(1-v)_N}\frac{e^{2\pi i\xi}\Gamma(1-v+N)}{\Gamma(u+N)}\left\{\int_1^\infty \frac{t^t(u+v-2)}{e^t - e^{2\pi i\xi}}\,dt\right.$$

$$\left. + \int_0^1\left(\frac{t}{e^t - e^{2\pi i\xi}} - \sum_{k=0}^{K}\frac{\beta_k(\xi)}{k!}t^k\right)t^{u+v-2}dt + \sum_{k=0}^{K}\frac{\beta_k(\xi)}{k!}\frac{1}{u+v+k-2}\right\}$$

$$- \int_0^1 \frac{e^{2\pi i\xi}}{\Gamma(u+N)}\left[\int_1^\infty \frac{e^{-xt}}{e^t - e^{2\pi i\xi}}t^{u+N-1}\,dt\right.$$

$$+ \int_0^1\left(\frac{t}{e^t - e^{2\pi i\xi}} - \sum_{k=0}^{K}\frac{\beta_k(\xi)}{k!}t^k\right)e^{-xt}t^{u+N-2}\,dt$$

$$\left. + \sum_{k=0}^{K}(\Gamma(u+N+k-1) - \Gamma(u+N+k-1,x))\frac{\beta_k(\xi)}{k!}x^{-u-N-k+1}\right]x^{N-v}dx$$

$$-\frac{(u)_N e^{2\pi i\xi}}{(1-v)_N}\sum_{k=0}^{K}\frac{\Gamma(u+N+k-1)}{\Gamma(u+N)k!}\frac{\beta_k(\xi)}{u+v+k-2}.$$

Using (3.3)′ and (3.5)′ for the first and the second integral, respectively, we conclude that

$$S_3 = \frac{(u)_N}{(1-v)_N}\left\{\frac{\Gamma(1-v+N)}{\Gamma(u+N)}\Gamma(u+v-1)\ell_{u+v-1}(\xi)\right. \tag{3.13′}$$

$$\left. - \int_0^1 e^{2\pi i\xi}\phi(\xi, x+1, u+N)x^{N-v}\,dx\right\}$$

$$-\frac{(u)_N e^{2\pi i\xi}}{(1-v)_N}\sum_{k=0}^{K}\frac{\Gamma(u+N+k-1)}{\Gamma(u+N)k!}\frac{\beta_k(\xi)}{u+v+k-2}.$$

We note that as in [SE, (2.12)], we have

$$\frac{(u)_N}{(1-v)_N}\int_0^1 e^{2\pi i\xi}\phi(\xi, x+1, u+N)x^{N-v}dx = T_N(u,v;\xi).$$

Thus, (3.13)′ gives

$$S_3 = \frac{\Gamma(u)}{\Gamma(1-v)}\Gamma(u+v-1)\ell_{u+v-1}(\xi) - T_N(u,v;\xi) \tag{3.14′}$$

$$-\frac{(u)_N e^{2\pi i\xi}}{(1-v)_N}\sum_{k=0}^{K}\frac{\Gamma(u+N+k-1)}{\Gamma(u+N)k!}\frac{\beta_k(\xi)}{u+v+k-2}.$$

Substituting (3.8)′ and (3.14)′ in (2.10)′ yields

$$\int_1^\infty g_K(\xi,u,v,x)\,dx = -S_N(u,v;\xi) - T_N(u,v;\xi) \tag{3.15′}$$

$$+\Gamma(u+v-1)\ell_{u+v-1}(\xi)\frac{\Gamma(u)}{\Gamma(1-v)}$$

$$-\sum_{k=0}^{\kappa}\frac{\Gamma(u+k-1)}{\Gamma(u)k!}\frac{e^{2\pi i\xi}\beta_k(\xi)}{u+v+k-2}.$$

Substituting (3.15)′ and its counterpart for $g_\kappa(-\xi, v, u, x)$ in

$$\int_1^2 f_\kappa(\xi,u,v,x)\,dx=\int_1^\infty -\triangle f_\kappa(\xi,u,v,x)\,dx \tag{2.14$'$}$$

$$=\int_1^\infty g_\kappa(\xi,u,v,x)\,dx+\int_1^\infty g_\kappa(-\xi,u,v,x)\,dx,$$

we conclude that

$$\int_1^2 f_\kappa(\xi,u,v,x)\,dx=-S_N(u,v;\xi)-S_N(v,u;-\xi) \tag{3.16$'$}$$

$$-T_N(u,v;\xi)-T_N(v,u;-\xi)+\left(\frac{\Gamma(u)}{\Gamma(1-v)}+\frac{\Gamma(v)}{\Gamma(1-u)}\right)\Gamma(u+v-1)\ell_{u+v-1}(\xi)$$

$$-\sum_{k=0}^{\kappa}\frac{\Gamma(u+k-1)}{\Gamma(u)k!}\frac{e^{2\pi i\xi}\beta_k(\xi)}{u+v+k-2}-\sum_{k=0}^{\kappa}\frac{\Gamma(v+k-1)}{\Gamma(v)k!}\frac{e^{-2\pi i\xi}\beta_k(\xi)}{u+v+k-2}.$$

To complete the proof of Theorem 3, it is enough to substitute (3.16)′ in

$$\int_1^\infty \phi(\xi,u,x)\phi(-\xi,v,x)dx=\int_1^2 f_\kappa(\xi,u,v,x)dx \tag{2.19$'$}$$

$$+\frac{1}{u+v-1}+\sum_{k=0}^{\kappa}\frac{\Gamma(u+k-1)}{\Gamma(u)k!}\frac{e^{2\pi i\xi}\beta_k(\xi)}{u+v+k-2}$$

$$+\sum_{k=0}^{\kappa}\frac{\Gamma(v+k-1)}{\Gamma(v)k!}\frac{e^{-2\pi i\xi}\beta_k(-\xi)}{u+v+k-2}.$$

References

[1] T.M. Apostol, On the Lerch zeta-function, *Pacific J. Math.* **1** (1951), 161–167.

[2] T.M. Apostol, Addendum to 'On the Lerch zeta-function', *Pacific J. Math.* **2** (1952), 10.

[3] S. Egami and K. Matsumoto, Asymptotic expansion of multiple zeta functions and power mean values of Hurwitz zeta functions, *J. London Math. Soc.* (2) **66** (2002), 41–60.

[4] A. Erdélyi, W. Magnus, F. Oberhettinger and F.G. Tricomi (The Bateman Manuscript Project), *Higher Transcendental Functions*, Vol. **I**, McGraw-Hill, New York, Toronto, and London, 1953.

[5] E. Grosswald, *Representations of integers as sums of two squares*, Springer Verlag, New York-Berlin-Heidelberg-Tokyo, 1985.

[6] S. Kanemitsu, Y. Tanigawa and M. Yoshimoto, Structural elucidation of the mean square of the Hurwitz zeta-function, (submitted for publication).

[7] S. Kanemitsu, Y. Tanigawa, H. Tsukada and M. Yoshimoto, Contribution to the theory of the Hurwitz zeta-function, (submitted for publication).

[8] M. Katsurada, An application of Mellin-Barnes' type integrals to mean square of Lerch zeta-functions, *Collect. Math.* **48** (1997), 137–153.

[9] M. Katsurada, Power series and asymptotic series associated with the Lerch zeta-function, *Proc. Japan Acad. Ser. A Math. Sci.* **74** (1998), 167–170.

[10] A. Laurinčikas and R. Garunkštis, *The Lerch zeta-function*, Kluwer Academic Publ., Dordrecht-Boston-London, 2002.

[11] M. Mikolás, A simple proof of the functional equation for the Riemann zeta-function and a formula of Hurwitz, *Acta Sci. Math. (Szeged)* **18** (1957), 261–263.

[12] J. Milnor, On polylogarithms, Hurwitz zeta-functions, and the Kubert identities, *Ensein. Math.* (2) **29** (1983), 281–322.

[13] F. Oberhettinger, A simple proof of the functional equation for the Riemann zeta-function and a formula of Hurwitz, *Pacific J. Math.* **6** (1956), 117–128.

[14] H.M. Srivastava and J. Choi, *Series Associated with the Zeta and Related Functions*, Kluwer Academic Publishers, Dordrecht, Boston, and London, 2001.

[15] K. Ueno and M. Nishizawa, Quantum groups an zeta-functions, in *Quantum Groups: Formalism and Applications*, Proc. of the Thirtieth Karpacz Winter School (Karpacz, 1994) (J. Lukierski *et al.*, Editors), pp. 115-126, Polish Sci. Publ. PWN, Warsaw, 1995.

R. Balasubramanian
The Institute of Mathematical Sciences
Chennai-600113.

email: balu@imsc.res.in

S. Kanemitsu and H. Tsukada
Graduate School of Advanced Technology
Kinki University
Iizuka, Fukuoka
820-8555, Japan

email: kanemitu@fuk.kindai.ac.jp

The Riemann Zeta Function and Related Themes – 2006, pp. 39–43

Irreducible polynomials over number fields

Jasbir S. Chahal and M. Ram Murty

Dedicated to Professor K. Ramachandra on his 70th birthday

Abstract

Let K be a number field with ring of integers O_K. If $f \in O_K[x]$ takes on irreducible values infinitely often, then either f is itself an irreducible polynomial or is the product of an irreducible polynomial and a linear polynomial. The second possibility can arise and leads us to the notion of a Mersenne prime in number fields.

In 1837, Dirichlet proved his famous theorem on primes in an arithmetic progression: *if a, b are coprime integers with b positive, the arithmetic progression*

$$\{a + bn | n = 1, 2, 3, \ldots\}$$

contains infinitely many primes.

Suppose that

$$f(x) = a_0 + a_1 x + \cdots + a_m x^m \qquad (a_m > 0) \tag{1}$$

is an irreducible polynomial over $\mathbb{Z}$ such that the set of values $f(\mathbb{N})$ has no common factor > 1. Then Dirichlet's theorem says that for $m = 1$, $f(n)$ assumes prime values for infinitely many n in $\mathbb{N} = \{1, 2, 3, \ldots\}$. In 1854, Buniakowski conjectured that for all $m \geq 1$, $f(x)$ represents primes infinitely often. Buniakowski's conjecture has not been proved or disproved so far for any polynomial of degree $m > 1$. Heuristic arguments, as well as some extensive numerical evidence, suggest that it is true.

It is easy to see that the converse of Buniakowski's conjecture is true, that is, if $f(x)$ is a polynomial of degree ≥ 1 and $f(n)$ is a prime for infinitely many values of n, then $f(x)$ is an irreducible polynomial. To see this let $f(x) = g(x)h(x)$ with $g(x), h(x)$ in $\mathbb{Z}[x]$. Since $f(n)$ is a prime infinitely often, one of $g(n)$ and $h(n)$, say $g(n) = \pm 1$ for infinitely many n. This cannot happen if $\deg g(x) \geq 1$, hence $f(x)$ is irreducible. Actually, there is a stronger converse to Buniakowski's conjecture (see Theorem 1, [5]), but we shall be concerned with the above converse in a more general setting. For this, suppose K is a number field, that is, a finite extension of $\mathbb{Q}$. Let O_K be its ring of integers. We would like to study to what extent the converse of Buniakowski's conjecture does hold. More precisely, if $f(x)$ is in $O_K[x]$ and $f(\alpha)$ is an irreducible element of O_K for infinitely many α in O_K, does the polynomial $f(x)$ have to be irreducible (over O_K)? In this paper we prove the following:

Theorem 1. *Suppose $f(x) \in O_K[x]$ with non-zero discriminant. If $f(x)$ represents an irreducible element of O_K infinitely often, then either $f(x)$ is an irreducible polynomial or $f(x) = \ell(x)p(x)$, where $p(x)$ is an irreducible polynomial and $\ell(x)$ is a linear factor.*

Date: January 30, 2004.
1991 *Mathematics Subject Classification.* Primary: 11R04, Secondary: 11R27, 11C08, 11D45.
Key words and phrases. irreducible polynomials, irreducible elements

Later, we will show that the non-vanishing of the discriminant is essential. We will also give examples where the second possibility, namely that $f(x) = \ell(x)p(x)$, arises.

For the proof we need two well-known theorems in number theory and an ingenious trick due to S. Pillai (see pp. 78-79 of [1]). To recall the theorems, let for a commutative ring A with $1 \neq 0$,

$$A^\times = \{u \in A | uv = 1 \text{ for some } v \text{ in } A\}$$

denote its group of units. The first theorem we need is Dirichlet's theorem on units.

Theorem. (Dirichlet) *For a number field K, let W_K denote the roots of unity in K and $r + 1$ the cardinality of the set of infinite places of K. Then W_K is finite and there are r elements $\epsilon_1, \ldots, \epsilon_r$ in $O_K^\times$ such that any u in $O_K^\times$ has a unique representation*

$$u = w\epsilon_1^{a_1} \cdots \epsilon_r^{a_r}, \tag{2}$$

where $w \in W_K$ and $a_j \in \mathbb{Z}$. In other words, as a group

$$O_K^\times \cong W_K \times \mathbb{Z}^r.$$

Call an element π of O_K, π not in $O_K^\times$, *irreducible* if π cannot be written in a non-trivial way as the product of two elements in O_K. That is, $\pi = \alpha\beta$ with α, β in O_K implies α or $\beta \in O_K^\times$. Let $N_{K/\mathbb{Q}} : K \to \mathbb{Q}$ be the norm from K to $\mathbb{Q}$.

The second theorem we need is Siegel's theorem on integral points on curves of positive genus (see [4, p. 217]). (See also Mordell [2, pp. 264-265], where the results are stated for the rational number field but apply to general number fields.)

Theorem. (Siegel) *Suppose $F(X, Y) \in O_K[X, Y]$ such that*

$$F(X, Y) = 0 \tag{3}$$

defines an affine curve (whose projective completion is) of positive genus. Then there are only finitely many points on (3) with coordinates in O_K.

Examples of such curves are hyperelliptic curves given by the equations of the form

$$AY^m = G(X), \tag{4}$$

where $m \geq 2$, $A \neq 0$ is in O_K and $G(X) \in O_K[X]$ is of degree ≥ 3 with no multiple roots.

We state two corollaries of this theorem for future reference:

Corollary 1. *The equation*

$$y^2 = f(x)$$

where $f(x)$ is a polynomial in $O_K[x]$ with at least three distinct roots, has only finitely many solutions with co-ordinates in O_K.

Corollary 2. *If $a, b, c, d \in O_K$, $ad \neq 0$, $b^2 - 4ac \neq 0$, $n \geq 3$, then the equation*

$$ay^2 + by + c = dx^n$$

has only finitely many solutions in O_K.

Now suppose $f(x) \in O_K[x]$. For $f(x)$ to be irreducible, all its roots must be simple. This necessary condition for irreducibility is easily checked because $f(x)$ has a repeated root if and only if the resultant $R(f, f')$ of $f(x)$ and its derivative $f'(x)$, which is computable, is zero (see [3, p. 203]). Thus we may assume that $f(x)$ has no repeated root, which we do.

Proof of Theorem 1. Suppose $f(x) = g(x)h(x)$ with $g(x)$, $h(x)$ in $O_K[x]$. If for infinitely many α in O_K,

$$\pi = f(\alpha) = g(\alpha)h(\alpha)$$

is irreducible, then one of $g(\alpha)$ and $h(\alpha)$, say $g(\alpha) \in O_K^\times$ infinitely often. Let

$$g(\alpha) = w(\alpha)\epsilon_1^{a_1(\alpha)} \cdots \epsilon_r^{a_r(\alpha)} \tag{5}$$

as in (2), where $w(\alpha)$ is a root of unity. We now use Pillai's trick. Write $a_j = a_j(\alpha) = 2b_j + c_j$, $c_j = 0$ or 1. Then (5) becomes

$$A\left(\epsilon_1^{b_1} \cdots \epsilon_r^{b_r}\right)^2 = g(\alpha), \tag{6}$$

where $A = w\epsilon_1^{c_1} \cdots \epsilon_r^{c_r}$. Hence for some choice of w and $c_1, \ldots, c_r$, (6) has infinitely many solutions in O_K. If $\deg g(x) \geq 3$ this cannot happen by Siegel's theorem (Corollary 1). Hence $\deg g(x) \leq 2$.

Next we write $a_j = 3b_j + c_j$ with $c_j = 0, 1, 2$. Then (5) becomes

$$w\epsilon_1^{c_1} \cdots \epsilon_r^{c_r}\left(\epsilon_1^{b_1} \cdots \epsilon_r^{b_r}\right)^3 = g(\alpha). \tag{7}$$

If $\deg g(\alpha) = 2$, we see that, with a slight change of variables, this leads to infinitely many solutions of (4) in O_K, where $m = 2$ and $\deg g(x) = 3$, which is again impossible by Corollary 2. Hence $\deg g(x) \leq 1$. This completes the proof of Theorem 1. □

We expect Theorem 1 to be the best possible. We say this in view of the following argument. Consider the quadratic field $K = \mathbb{Q}(\sqrt{D})$, $D > 0$. Let $f(x) = x(x-1)$. If ϵ is the fundamental unit of K, then $f(\epsilon^n) = \epsilon^n(\epsilon^n - 1)$. Thus Theorem 1 would be the best possible if $\pi_n = \epsilon^n - 1$ is an irreducible element of O_K for infinitely many n.

First note that $\pi = \epsilon^n - 1$ is a prime element of O_K if and only if $|N_{K/\mathbb{Q}}(\pi)|$ is a rational prime. To check the primality of $N_{K/\mathbb{Q}}(\pi)$, the following lemma is useful.

Lemma 1. *Let ϵ be the fundamental unit of $K = \mathbb{Q}(\sqrt{D})$, $D > 0$ such that $N(\epsilon) = -1$. If for n odd,*

$$\epsilon^n = \frac{a + b\sqrt{D}}{2},$$

where a or b is odd, then $N(\epsilon^n - 1) = -a$.

Proof:
$$\begin{aligned} N_{K/\mathbb{Q}}(\epsilon^n - 1) &= N_{K/\mathbb{Q}}\left(\frac{a + b\sqrt{D}}{2} - 1\right) \\ &= \frac{(a-2)^2 - Db^2}{4} \\ &= \frac{(a^2 - Db^2) - 4a + 4}{4} = -a \end{aligned}$$

since $\dfrac{a^2 - Db^2}{4} = N(\epsilon^n) = (-1)^n = -1$. □

For $\epsilon^n - 1 = (\epsilon - 1)(\epsilon^{n-1} + \cdots + \epsilon + 1)$ to be an irreducible element of O_K, one of its factors, which most likely is $\epsilon - 1$, must be a unit. Since $0 < \epsilon - 1 < \epsilon$, this requires that $\epsilon - 1 = \epsilon^{-r}$ with $r > 0$, that is, $\epsilon^r(\epsilon - 1) = 1$, which implies that $\epsilon - 1 < 1$. But $\epsilon < 2$ for the quadratic field $K = \mathbb{Q}(\sqrt{5})$. So let $K = \mathbb{Q}(\sqrt{5})$.

By Theorem 1, for $\epsilon^n - 1 = (\epsilon - 1)(\epsilon^{n-1} + \cdots + \epsilon + 1)$ to be an irreducible element of O_K infinitely often, $x^{n-1} + \cdots + x + 1$ has to be irreducible, hence n must be a prime. Let for a prime p, $\Phi_p(x)$ denote the cyclotomic polynomial $x^{p-1} + \cdots + x + 1$.

For the quadratic field $K = \mathbb{Q}(\sqrt{5})$, $\epsilon = \dfrac{1 + \sqrt{5}}{2}$ is the fundamental unit with $N_{K/\mathbb{Q}}(\epsilon) = -1$. Since $\epsilon - 1$ is the reciprocal of ϵ, we see that it is a unit. Table 1 shows the first six of the primes $p < 100$ for which $\Phi_p(\epsilon)$ is an irreducible element of O_K (cf. Lemma 1). The remaining ones are 31, 37, 41, 47, 53, 61, 71 and 79.

Table 1

p	ϵ^p	$\lvert N_{K/\mathbb{Q}}(\epsilon^p - 1)\rvert$
5	$\dfrac{11 + 5\sqrt{5}}{2}$	11 (prime)
7	$\dfrac{29 + 13\sqrt{5}}{2}$	29 (prime)
11	$\dfrac{199 + 89\sqrt{5}}{2}$	199 (prime)
13	$\dfrac{521 + 233\sqrt{5}}{2}$	521 (prime)
17	$\dfrac{3571 + 1597\sqrt{5}}{2}$	3571 (prime)
19	$\dfrac{9349 + 4181\sqrt{5}}{2}$	9349 (prime)
23	$\dfrac{64079 + 28657\sqrt{5}}{2}$	$64079 = 139 \cdot 461$

It is reasonable to expect that $\Phi_p(\epsilon)$ is a prime element of O_K for infinitely many rational primes p, providing a strong evidence that Theorem 1 is the best possible. We remark that the same reasoning with the polynomial $f(x) = x^2(x - 1)$ suffices to show that the condition of non-vanishing discriminant is also essential in our theorem.

References

[1] G.H. Hardy, *Ramanujan, Twelve Lectures on Subjects Suggested by His Life and Work*, Chelsea Publishing Company, New York, N.Y., 3rd Edition, 1978.

[2] L.J. Mordell, *Diophantine Equations*, Academic Press (1969).

[3] Serge Lang, *Algebra* (3rd Ed.), Addison-Wesley (1999).

[4] Serge Lang, *Survey of Diophantine Geometry*, Springer Verlag (1997).

[5] M. Ram Murty, *Prime Numbers and Irreducible Polynomials*, American Math. Monthly, **109** (2002), pp. 452–458.

Jasbir S. Chahal
Department of Mathematics,
Brigham Young University,
Provo, UT 84602, USA.

e-mail: jasbir@math.byu.edu

M. Ram Murty
Department of Mathematics,
Queen's University, Kingston,
Ontario, K7L 3N6, Canada.

e-mail: murty@mast.queensu.ca

The Riemann Zeta Function and Related Themes – 2006, pp. 45–56

On a conjecture of Andrews - III

Chandrashekara B.M., Padmavathamma and Raghavendra, R.

Dedicated to Professor K. Ramachandra on his 70th birthday

Abstract

$A_{\lambda,k,a}(n)$ denotes the number of partitions of n subject to certain restrictions on their parts and $B_{\lambda,k,a}(n)$ denotes the number of partitions of n subject to certain other restrictions on their parts, both too long to be stated in the abstract. In 1974 Andrews [On the General Rogers-Ramanujan theorem, Mem.Amer.Math. Soc.No.152,(1974), 1-86, Conjecture-1] stated a conjecture on these two partition functions. The case $k = a$ was proved by the first author and T.G.Sudha [on a conjecture of Andrews, Inter.J.Math and Math.Sci, Vol.16, No.4, 1993, 763-774]. The first author and Ruby Salestina.M [on a conjecture of Andrews-II, Number Theory and Discrete Mathematics, Proc.International Conference in Honour of Srinivasa Ramanujan, Chandigarh, India, 2000, 135-147] have established the case $k = a + 1$ and proved [1974 conjecture of Andrews on partitions, Inter.J.Math and Math.sci (to appear)] that the conjecture is false if n exceeds $(2k - a - \frac{\lambda}{2} + 1)(\lambda + 1)$ for even λ and $(4k - 2a - \lambda + 2)(\frac{\lambda+1}{2})$ for odd λ and $k \geq a + 2$. In that paper they had also stated a revised conjecture. The object of this paper is to give a proof.

1 Introduction

For an even integer λ, let $A_{\lambda,k,a}(n)$ denote the number of partitions of n such that

- no part $\not\equiv 0 \pmod{\lambda + 1}$ may be repeated, and
- no part is $\equiv 0, \pm\left(a - \frac{\lambda}{2}\right)(\lambda + 1) \pmod{(2k - \lambda + 1)(\lambda + 1)}$.

For an odd integer λ, let $A_{\lambda,k,a}(n)$ denote the number of partitions of n such that

- no part $\not\equiv 0 \pmod{\frac{\lambda+1}{2}}$ may be repeated,
- no part is $\equiv \lambda + 1 \pmod{2\lambda + 2}$, and
- no part is $\equiv 0, \pm(2a - \lambda)\left(\frac{\lambda+1}{2}\right) \pmod{(2k - \lambda + 1)(\lambda + 1)}$.

Let $B_{\lambda,k,a}(n)$ denote the number of partitions of n of the form $b_1 + \cdots + b_s$ with $b_i \geq b_{i+1}$, such that

- no part $\not\equiv 0 \pmod{\lambda + 1}$ is repeated,
- $b_i - b_{i+k-1} \geq \lambda + 1$, with strict inequality if b_i is a multiple of $\lambda + 1$, and

---

2000 *Mathematics Subject Classification.* Primary 11P81, 11P82, 11P83; Secondary 05A15, 05A17, 05A19.
Key words and phrases. Partition Functions, Generating Function, Andrews Conjecture
The first author is partially and the second and third authors are fully financially supported by Department of Science and Technology, New Delhi, India, grant DST/MS/116/2k, "Some Studies in Partition Theory".

- $\sum_{i=j}^{\lambda-j+1} f_i \leq a-j$ for $1 \leq j \leq \frac{\lambda+1}{2}$ and $f_1 + \cdots + f_{\lambda+1} \leq a-1$, where f_j is the number of appearances of j in the partition.

Andrews [3] proved a general theorem from which the well-known Rogers-Ramanujan identities, Gordon's theorem [7], the Gollnitz-Gordon identities [6] and their generalization [1], Schur's theorem and its generalization [14] could be deduced. The following theorem was proved by Andrews [2] in 1969.

Theorem 1 (2, Th.2). *If λ, k, and a are positive integers with $\frac{\lambda}{2} \leq a \leq k$, $k \geq 2\lambda - 1$ then for every positive integer, we have*

$$A_{\lambda,k,a}(n) = B_{\lambda,k,a}(n)$$

Schur's theorem [14] is the case $\lambda = k = a = 2$. Hence it is not a particular case of Theorem 1 as $k \geq 2\lambda - 1$ is not satisfied. This lead Andrews [2] to conjecture that Theorem 1 may be still true if $k \geq \lambda$. In fact he [4] gave a proof of this result. Andrews [4] stated the following two conjectures.

Conejecture 1. *For $\frac{\lambda}{2} < a \leq k < \lambda$, let $n^c = \frac{(k+\lambda-a+1)(k+\lambda-a)}{2} + (k-\lambda+1)(\lambda+1)$. Then*

$$\begin{aligned} B_{\lambda,k,a}(n) &= A_{\lambda,k,a}(n) \quad &\text{for} \quad 0 \leq n < n^c \\ \text{and} \quad B_{\lambda,k,a}(n) &= A_{\lambda,k,a}(n) + 1 \quad &\text{for} \quad n = n^c. \end{aligned}$$

Conejecture 2. *There holds the identity $A_{4,3,3}(n) = B^0_{4,3,3}(n)$ for all positive integers n, where $B^0_{4,3,3}(n)$ denotes the number of partitions of n enumerated by $B_{4,3,3}(n)$ with the added restrictions:*

$$\begin{aligned} f_{5j+2} + f_{5j+3} &\leq 1 \quad \text{for} \quad j \geq 0, \\ f_{5j+4} + f_{5j+6} &\leq 1 \quad \text{for} \quad j \geq 0, \\ f_{5j-1} + f_{5j} + f_{5j+5} + f_{5j+6} &\leq 3 \quad \text{for} \quad j \geq 1, \end{aligned}$$

where, as before, f_j denotes the number of appearances of j in the partition.

Conjecture 2 is designed to show that some partition identities can be obtained in a few cases when the condition $k \geq \lambda$ is removed with some additional restrictions on the summands. In the year 1994 Andrews et al [5] gave an analytical proof of conjecture 2. The first author and Ruby Salestina,M [11] gave a combinatorial proof. These two authors and Sudarshan, S.R [12] first conjectured and then proved combinatorially the following result which is analogous to conjecture 2.

Theorem 2. *There holds the identity $A_{5,3,3}(n) = B^0_{5,3,3}(n)$ for all positive integers n, where $B^0_{5,3,3}(n)$ denotes the number of partitions of n enumerated by $B_{5,3,3}(n)$ with the added restrictions:*

$$\begin{aligned} f_{6j+3} &= 0 \quad \text{for} \quad j \geq 0, \\ f_{6j+2} + f_{6j+4} &\leq 1 \quad \text{for} \quad j \geq 0, \\ f_{6j+5} + f_{6j+7} &\leq 1 \quad \text{for} \quad j \geq 0, \\ f_{6j-1} + f_{6j} + f_{6j+6} + f_{6j+7} &\leq 3 \quad \text{for} \quad j \geq 1. \end{aligned}$$

In [13] the authors have given an analytic proof of Theorem 2. The first author and T.G. Sudha[8] have proved the case $k = a$ of conjecture 1. The first author and Ruby Salestina.M [9] have established the case $k = a + 1$ and proved [10] that the conjecture is false if n exceeds $(2k - a - \frac{\lambda}{2} + 1)(\lambda + 1)$ for even λ and $(4k - 2a - \lambda + 2)(\frac{\lambda+1}{2})$ for odd λ and $k \geq a + 2$ by giving counter examples. They had also stated the following revised conjecture for a particular case when λ is even.

Conejecture 3 (Revised). *Let λ be even, $a - \frac{\lambda}{2} = 1$, $\theta = k - a$, $\frac{\theta(\theta-1)}{2} < \left(a - \frac{\lambda}{2}\right)(\lambda + 1)$ and $0 \leq \theta \leq \frac{\lambda}{2} - 3$. Then*

$$B_{\lambda,k,a}(n) = A_{\lambda,k,a}(n) \quad \textit{for} \quad n < (2k - a - \frac{\lambda}{2} + 1)(\lambda + 1) \tag{1}$$

$$B_{\lambda,k,a}(n) = A_{\lambda,k,a}(n) + B_{\lambda,k,a}(x)$$
$$\textit{where } n = (2k - a - \frac{\lambda}{2} + 1)(\lambda + 1) + x, 0 \leq x \leq \frac{\theta(\theta - 1)}{2}. \tag{2}$$

The object of this paper is to give a proof of this revised conjecture.

2 Preliminaries

Let $P_{B_{\lambda,k,a}}(n)$ and $P_{A_{\lambda,k,a}}(n)$ denote the set of partitions enumerated by $B_{\lambda,k,a}(n)$ and $A_{\lambda,k,a}(n)$ respectively. Let $P'_A(n)$ [resp. $P'_B(n)$] denote the set of partitions enumerated by $A_{\lambda,k,a}(n)$ [resp. $B_{\lambda,k,a}(n)$] but not by $B_{\lambda,k,a}(n)$ [resp. $A_{\lambda,k,a}(n)$].

$\pi \in P'_A(n)$ implies that it violates one of the conditions on $f's$ or $b's$. Let $S_j(j = 1, 2, ldots, \frac{\lambda}{2})$ denote the condition $\sum_{i=j}^{\lambda-j+1} f_i \leq a - j$ and let S denote the condition $\sum_{i=1}^{\lambda+1} f_i \leq a - 1$ and let S^* be the condition on $b's$.

Let $(2k - a - \frac{\lambda}{2} + 1)(\lambda + 1) \leq n < (2k - a - \frac{\lambda}{2} + 1)(\lambda + 1) + \frac{\theta(\theta-1)}{2}$ where $\frac{\theta(\theta-1)}{2} < (a - \frac{\lambda}{2})(\lambda + 1)$ and $\theta = k - a$. Then

$$P'_B(n) = Q^1 \cup \cdots \cup Q^{a-1} \cup R(n)$$

where for $1 \leq i \leq a - 1$,

$$Q^i = \left\{\pi \in P'_B(n) : (a - \frac{\lambda}{2})(\lambda + 1) \text{ appears } i \text{ times}\right\} \text{ and}$$
$$R(n) = \{(2k - a - (\lambda/2) + 1)(\lambda + 1) + \pi : \pi \text{ is a partition of}$$
$$n - (2k - a - (\lambda/2) + 1)(\lambda + 1) \text{ into parts with } C\}$$

Here C stands for "subjected to the conditions in the definition of B".

Clearly # $R(n) = B_{\lambda,k,a}[n - (2k - a - \frac{\lambda}{2} + 1)(\lambda + 1)]$.

From the method explained in [8] and [9] it follows that the partitions violating $S_1, \ldots, S_{\frac{\lambda}{2}}$ will be mapped onto $Q^1 \cup \cdots \cup Q^{a-1}$. If $a - \frac{\lambda}{2} = 1$ then S reduces to S_1. As such any contribution to $R(n)$ can come only from those partitions of P'_A which violate S^* but do not violate any of $S_1, \ldots, S_{\frac{\lambda}{2}}$. If there are no partitions of n violating only S^* then for such n, we have

$$P'_A(n) = \text{ Union of the partitions violating } S_1, \ldots, S_{\frac{\lambda}{2}}$$

3 Proof of The Revised Conjecture

Let λ be even. $\pi \in P'_A(n)$ implies that it violates one of the conditions on f's or b's. In [8] and [9] we have shown that for $n < (2k-a-\frac{\lambda}{2}+1)(\lambda+1)$, if a partition violates S* then it violates S or S_1. However if $n \geq (2k-a-\frac{\lambda}{2}+1)(\lambda+1)$ then there exist partitions which violate S* but do not violate any of $S, S_1, ldots, S_{\frac{\lambda}{2}}$. For example when $\lambda = 14$, $k = 13$, $a = 8$, $\theta = 5$, $(2k-a-\frac{\lambda}{2}+1)(\lambda+1) = 180$ then n in conjecture 2 is 190.

$$21 + \cdots + 16 + 14 + \cdots + 9 + 7$$
$$21 + \cdots + 16 + 14 + \cdots + 9 + 8$$

are the partitions of 187 and 188 respectively which violate only S*.
Let us now investigate such partitions.
If a partition violates S* then there exists a partition

$$n = b_1 + \cdots + b_i + \cdots + b_{i+k-1} + \cdots + b_k + \cdots + b_s \tag{3}$$

and an integer i with $b_i - b_{i+k-1} < \lambda + 1$. If $b_{i+k-1} \geq \lambda + 1$ then the number being partitioned is

$$\begin{aligned} &\geq (\lambda + x_k) + \cdots + (\lambda + x_1) + \cdots \\ &\geq k(\lambda+1) \ \text{ where } \ x_k - x_1 < \lambda + 1 \end{aligned} \tag{4}$$

If (4) contains $\lambda + 1$ more than $(a-1)$ times then it violates S. Let x denote the number of $\lambda+1$ in (4) and y denote the number of terms $> \lambda+1$ so that $x \leq a-1$ and $x + y = k$. Then (4) becomes

$$x(\lambda+1) + (\lambda+2) + \cdots + (\lambda+k-x) = (k-1)(\lambda+1) + \frac{(k-x)(k-x-1)}{2}$$

Let n^c denote the n in the conjecture. If $k = a + \theta$ then for $0 \leq x' \leq \frac{\theta(\theta-1)}{2}$, we have $n = (2k-a-\frac{\lambda}{2}+1)(\lambda+1) + x'$

$$\begin{aligned} \leq n^c &= (2k-a-\frac{\lambda}{2}+1)(\lambda+1) + \frac{\theta(\theta-1)}{2} \\ &= k(\lambda+1) + (k-a-\frac{\lambda}{2}+1)(\lambda+1) + \frac{(k-a)(k-a-1)}{2} \\ &< k(\lambda+1) + (k-a-\frac{\lambda}{2}+1)(\lambda+1) + \frac{(k-x)(k-x-1)}{2} \\ &\leq (k-1)(\lambda+1) + \frac{(k-x)(k-x-1)}{2} \qquad \text{since } k-a-\frac{\lambda}{2}+1 < 0. \end{aligned}$$

Let $b_{i+k-1} < \lambda+1$ and $b_i < \lambda+1$. Then (3) contains at least k parts $\leq \lambda$ and hence $\sum_{i=1}^{\lambda} f_i \geq k > a-1$ which implies that such a partition violates S_1.

Let $b_{i+k-1} < \lambda+1$ and $b_i \geq \lambda+1$. If the number of parts among $1, 2, \ldots, \lambda+1$ is $\geq a$ then the partition violates S or S_1. Let β denote the number of parts among $1, 2, \ldots, \lambda+1$. Then $1 \leq \beta \leq a-1$. Let α denote the number of parts $> \lambda+1$ so that $k-a+1 \leq \alpha \leq k-1$. Then the number being partitioned is

$$(\lambda + x_\alpha) + \cdots + (\lambda + x_1) + y_1 + \cdots + y_\beta. \tag{5}$$

Since $\lambda + x_\alpha - y_\beta < \lambda + 1$, we have $x_\alpha = y_\beta$. Now $x_1 \geq 2, x_2 \geq 3, \ldots, x_\alpha \geq \alpha + 1$. Thus $y_\beta \geq \alpha + 1, \ldots, y_1 \geq \alpha + \beta = k$. Hence (5) is

$$\geq (\lambda + \alpha + 1) + \cdots + (\lambda + 2) + (\alpha + \beta) + \cdots + (\alpha + 1)$$
$$= \alpha(\lambda + 1) + \frac{(\alpha + \beta)(\alpha + \beta + 1)}{2}$$

i) Let $\beta = 1$. Then (5) becomes

$$(k-1)(\lambda+1) + \frac{k(k+1)}{2} > (2k - a - \frac{\lambda}{2} + 1)(\lambda + 1) + \frac{\theta(\theta - 1)}{2} = n^c$$

for their difference is

$$= \left[\frac{(\lambda - 2)}{2} - \theta\right](\lambda + 1) + \frac{(k+\theta)(a+1)}{2} > 0$$

since $0 \leq \theta \leq \frac{\lambda}{2} - 2$ and $k = a + \theta$.

Proceeding like this we arrive at the $(a-1)^{th}$ step.

a-1) Let $\beta = a - 1$. Then $\alpha = k - a + 1$. Let

$$S_1^* = \{k - a + 2, \ldots, \lambda - k + a - 1\}$$
$$\text{and} \quad S_2^* = \{\lambda - k + a, \ldots, \lambda + 1\}$$

The number of terms in S_1^* and S_2^* are respectively $\lambda - 2k + 2a - 2$ and $k - a + 2$. Since we have to choose $(a-1)$ parts from $\{1, 2, \ldots, \lambda + 1\}$ and $k - a + 1$ parts $> \lambda + 1$ for a partition violating S^*; it is clear that the minimum part should be $k - a + 2$. Hence we consider the condition

$$S_{k-a+2} : f_{k-a+2} + \cdots + f_{\lambda-k+a-1} \leq a - (k - a + 2) = 2a - k - 2$$

If the number x of terms in S_1^* satisfies $(2a - k - 2) < x \leq (\lambda - 2k + 2a - 2)$ and the number y of terms in S_2^* satisfies $x + y = a - 1$ then the partition violates S_{k-a+2}. Since the number y of terms in S_2^* is $k - a + 2$ and we have to choose $a - 1$ terms from S_1^* and S_2^*, the minimum number in S_1^* is $(a-1) - (k - a + 2) = 2a - k - 3$. Thus we are left with two choices for x namely $x = 2a - k - 3$ and $x = 2a - k - 2$.

In case of $S_{k-a+3}, \ldots, S_{\frac{\lambda}{2}}$

$$S_{k-a-3} : f_{k-a+3} + \cdots + f_{\lambda-k+a} \leq a - (k - a + 3) = 2a - k - 3$$
$$S_1^* = \{k - a + 3, \ldots, \lambda - k + a\} \quad \# S_1^* = \lambda - 2k + 2a - 2$$
$$S_2^* = \{\lambda - k + a + 1, \ldots, \lambda + 1\} \quad \# S_2^* = k - a + 1$$

$$2a - k - 3 < x \leq \lambda - 2k + 2a - 2$$

$(\lambda + 1) - (k - a + 3) + 1 = \lambda - k + a - 1 \rightarrow$ The total number of terms in S_1^* and S_2^*. Number of terms in S_2^* is $k - a + 1$ and we have to choose $a - 1$ terms from S_1^* and S_2^*, the minimum number in S_1^* is $(a-1) - (k - a + 1) = 2a - k - 2 \Rightarrow x > 2a - k - 3 \Rightarrow$ it violates S_{k-a+3}. Similarly we can say for S_{k-a+4} and so on.

Let $x = (2a - k - 3)$ and let

$$\begin{aligned} A &= \{(k-a+2) + \cdots + [(k-a+2) + (2a-k-4)]\} \\ &\quad + \{(\lambda - k + a) + \cdots + [(\lambda - k + a) + (k - a + 1)]\} \\ &= (k-a+2)(2a-k-3) + \frac{(2a-k-4)(2a-k-3)}{2} \\ &\quad + (\lambda - k + a)(k - a + 2) + \frac{(k-a+1)(k-a+2)}{2} \\ \text{and} \quad B &= \{(\lambda + 1 + 1) + \cdots + [(\lambda + 1) + (k - a + 1)]\} \\ &= (\lambda + 1)(k - a + 1) + \frac{(k-a+1)(k-a+2)}{2} \end{aligned}$$

Then

$$A + B - n^c = 2ak + \frac{\lambda^2}{2} + \frac{5\lambda}{2} + 2 - k^2 - \frac{a^2}{2} - k - \lambda a - \frac{3a}{2} \tag{6}$$

Let $x = 2a - k - 2$.

Then analogous to (6) we get

$$A^* + B^* - n^c = 2ak + \frac{\lambda^2}{2} + \frac{5\lambda}{2} + 2 - k^2 - \frac{a^2}{2} - k - \lambda a - \frac{3a}{2} - (\lambda - a + 2). \tag{7}$$

Lemma 1. *Let* $a - \frac{\lambda}{2} = 1$ *and let* $\frac{\theta(\theta-1)}{2} < (\lambda + 1)$. *For* $0 \le \theta \le \frac{\lambda}{2} - 3$ *there are no partitions of n violating only* S^*.

Proof: Putting $k = a + \theta$, (7) reduces to

$$A^* + B* - n^c = (\lambda - a)(\lambda - a + 3) - 2\theta(\theta + 1) \tag{8}$$

For $\theta = 0, 1, 2$, (8) > 0. Proceeding like this we arrive at the value of $\theta = \frac{\lambda}{2} - 3$.

Now consider,

$$\frac{\theta(\theta-1)}{2} - (\lambda + 1) = \frac{1}{8}(\lambda^2 - 22\lambda + 40) \ge 0 \text{ for } \lambda \ge 20$$

Hence, when $\theta = \frac{\lambda}{2} - 3$ we have $\frac{\theta(\theta-1)}{2} < (\lambda + 1)$ for $\lambda \le 18$. But it is easy to see that (8) > 0 for $\lambda \le 18$, $\theta = \frac{\lambda}{2} - 3$, and $k = \lambda - 2$. This proves Lemma 1. □

Lemma 2. *Cardinality of* $Q^1 \cup Q^2 \cup \cdots \cup Q^{a-1}$ = *Cardinality of* $P'_A(n)$. *under the conditions of the Revised conjecture.*

Proof: $\pi \epsilon P'_A(n)$ implies that it violates one of the conditions $S_1, \ldots, S_{\frac{\lambda}{2}}, S, S^*$. Since $a - \frac{\lambda}{2} = 1$, S reduces to S_1. In Lemma 1, we have proved that there are no partitions of n violating only S^*. Thus $\pi \epsilon P'_A(n)$ implies that it violates one of the conditions $S_1, \ldots, S_{\frac{\lambda}{2}}$. We now give the bijection from $P'_A(n)$ onto the set $X_B(n)$ of partitions enumerated by $Q^1 \cup Q^2 \cup \cdots \cup Q^{a-1}$. □

Bijection from $P'_A(n)$ onto $X_B(n)$

Definition 1. A pair (α,β) shall be called a P-pair if $\alpha < \beta$ and $\alpha + \beta = \lambda + 1$.

Definition 2. We say that a P-pair (α_1,β_1) in a partition π is Connected to another P-pair (α_2,β_2) if

$$\alpha_1 < \alpha_2 \text{ and } (\alpha_1 + 1 \text{ or } \beta_1 - 1) \text{ and } (\alpha_1 + 2 \text{ or } \beta_1 - 2)\cdots \text{ and } \\ (\alpha_2 - 1 \text{ or } \beta_2 + 1) \text{ are present in } \pi. \tag{9}$$

Here onwards all the examples will be illustrated for the following values of λ, k, a.

$$\lambda = 14, k = 12, a = 8. \tag{10}$$

Here $\theta = k - a = 4$ satisfies

$$\frac{\theta(\theta-1)}{2} = \frac{4*3}{2} = 6 < (a - \frac{\lambda}{2})(\lambda + 1) = 15$$

and $0 \le \theta \le \frac{\lambda}{2} - 3 = 7 - 3 = 4$ and hence conditions in the Revised conjecture are satisfied.

The conditions $S, S_1, \ldots, S_7$ are -

$$\begin{aligned}
S &: f_1 + f_2 + \cdots + f_{14} + f_{15} \le 7\\
S_1 &: f_1 + f_2 + \cdots + f_{13} + f_{14} \le 7\\
S_2 &: f_2 + f_3 + \cdots + f_{12} + f_{13} \le 6\\
S_3 &: f_3 + f_4 + \cdots + f_{11} + f_{12} \le 5\\
S_4 &: f_4 + f_5 + \cdots + f_{10} + f_{11} \le 4\\
S_5 &: f_5 + f_6 + \cdots + f_9 + f_{10} \le 3\\
S_6 &: f_6 + f_7 + f_8 + f_9 \le 2\\
S_7 &: f_7 + f_8 \le 1
\end{aligned}$$

e.g., In a partition π, the P-pair (2,13) is connected to the P-pair (5,10) if (3 or 12) and (4 or 11) are parts of π.

Mapping from $P'_A(n)$ to $X_B(n)$

Let $\pi \epsilon P'_A(n)$. Find the greatest α say α_1 in π such that S_{α_1} is violated. This implies that α_1 and $\lambda + 1 - \alpha_1 = \beta_1$ are parts in π. Including the P-pair (α_1,β_1) replace all the P-pairs connected to (α_1,β_1) with $\lambda + 1$. Let the resulting partition be π'.

Let $\pi' = (\lambda + 1)j + \beta_t + \cdots + \alpha_k$.

Let α_m be the least among the replaced (with $\lambda + 1$) P-pairs. Then look for the greatest α_i ($\alpha_i < \alpha_m$) such that the number of elements from α_i to β_i is $\ge (a - \alpha_i + 1 - j)$. Including the P-pair (α_i,β_i) replace all the P-pairs connected to (α_i,β_i). Repeat the procedure as long as we get such P-pair (α_i,β_i). The Resulting partition $\psi \epsilon X_B(n)$.

e.g.1, $\pi = 14 + 13 + 12 + 11 + 10 + 8 + 7 + 5 + 4 + 3 + 2 + 1$.

Here S_7 is violated. There are no P-pairs connected to (7,8) since neither 6 nor 9 is a part in π. Thus we replace only (7,8).

$$\pi \to \pi' = 15 + 14 + 13 + 12 + 11 + 10 + 5 + 4 + 3 + 2 + 1.$$

Here $j = 1$. 7 is the least α among the replaced P-pairs. And 4 is the greatest α $(4 < 7)$, such that the number of elements from 4 to 11 is $4 \geq 4(= a - 4 + 1 - j)$. The P-pairs (3,12), (2,13), (1,14) are all connected to (4,11). Hence we replace them including (4,11) by 15. The resulting partition is

$$\psi = 15 + 15 + 15 + 15 + 15 + 10 + 5$$

we associate π to $\psi \epsilon X_B(n)$.

e.g.2, $\pi = 14 + 11 + 10 + 8 + 7 + 4 + 1$.

Here S_7 is violated. There are no P-pairs connected to (7,8) since neither 6 nor 9 is a part in π. Thus we replace only (7,8).

$$\pi \to \pi' = 15 + 14 + 11 + 10 + 4 + 1.$$

Here $j = 1$. 7 is the least α among the replaced P-pairs. There is no $\alpha < 7$ in the partition such that the number of elements between α to β is $\geq (a - \alpha + 1 - j)$. So we stop here. The resulting partition is

$$\psi = 15 + 14 + 11 + 10 + 4 + 1.$$

we associate π to $\psi \epsilon X_B(n)$.

e.g.3, $\pi = 12 + 11 + 10 + 5 + 4 + 3$.

Here S_3 is violated. There are no P-pairs connected to (3,12). Thus we replace only (3,12).

Since there is no P-pair (α, β) present such that $(\alpha < 3)$ we stop here and the resulting partition

$$\psi = 15 + 11 + 10 + 5 + 4 \epsilon X_B(n).$$

e.g.4, $\pi = 13 + 12 + 10 + 9 + 8 + 7 + 6 + 5 + 3 + 2$.

Here S_7 is violated. The P-pairs (6,9), (5,10) are connected to (7,8). Therefore

$$\pi \to \pi' = 15 + 15 + 15 + 13 + 12 + 3 + 2.$$

Here $j = 3$. 5 is the least α among the added P-pairs. And 2 is the greatest $\alpha (2 < 5)$, such that the number of elements from 2 to 13 is $4 \geq 4(= a - 2 + 1 - j)$. Hence we replace (2,13) by 15. The resulting partition is

$$\psi = 15 + 15 + 15 + 15 + 12 + 3 \epsilon X_B(n).$$

Reverse Mapping from $X_B(n)$ to $P'_A(n)$

Let $\psi \epsilon X_B(n)$. List the P-pairs (α, β) vertically one by one

$$\begin{array}{c} (\alpha_1, \beta_1) \\ (\alpha_2, \beta_2) \\ \cdots \\ \cdots \\ \cdots \\ (\alpha_n, \beta_n) \end{array}$$

where $\alpha_i < \alpha_j$ for $i < j$ and neither α_i nor β_i for $1 \leq i \leq n$ is a part in ψ.

Strike out the P-pair (α_k,β_k) from the list if there is a P-pair connected to (α_k,β_k) in the given partition ψ. Strike out the P-pairs (α_k,β_k) and $(\alpha_{k+1},\beta_{k+1})$ from the list if there are two P-pairs connected to (α_k,β_k). Likewise strike out the P-pairs (α_k,β_k) to $(\alpha_{k+l},\beta_{k+l})$ from the list if there are $l + 1$ P-pairs connected to (α_k,β_k).

Let the number of $(\lambda + 1)$ in ψ be j. Starting with the jth P-pair from the top of the list look for the P-pair (α,β) such that the replacement of that P-pair in place of a $(\lambda + 1)$ would violate the condition S_α.

We replace j P-pairs which are immediately above the P-pair (α,β) including the P-pair (α,β) with j times $(\lambda + 1)$. The resulting partition $\pi \epsilon P'_A(n)$.

e.g.1, $\psi = 15 + 15 + 15 + 15 + 15 + 10 + 5$. Here $j = 5$.

We list the P-pairs (α,β) vertically one by one where neither α nor β is a part in ψ.

$$\begin{array}{c}(1,14)\\(2,13)\\(3,12)\\(4,11)\\(6,9)\\(7,8)\end{array}$$

Since there is a P-pair (5,10) in ψ which is connected to the P-pair (6,9) in the list we strike out the P-pair (6,9).

Thus we are left with the P-pairs

$$\begin{array}{c}(1,14)\\(2,13)\\(3,12)\\(4,11)\\(7,8)\end{array}$$

From the top the $5^{th}(j = 5)$ pair is (7,8) and it violates S_7. Hence we replace five 15's in the partition by the pairs

$$(1,14),(2,13),(3,12),(4,11),(7,8)$$

The resulting partition is

$$\pi = 14 + 13 + 12 + 11 + 10 + 8 + 7 + 5 + 4 + 3 + 2 + 1$$

we associate ψ to $\pi \epsilon P'_A(n)$.

e.g.2, $\psi = 15 + 14 + 11 + 10 + 4 + 1$ Here $j = 1$.

We list the P-pairs (α,β) vertically one by one

$$\begin{array}{c}(2,13)\\(3,12)\\(6,9)\\(7,8)\end{array}$$

Since there is a P-pair (4,11) in ψ which is connected to the P-pair (6,9) in the list we strike out the P-pair (6,9) and since there is a P-pair (1,14) in ψ which is connected to the P-pair (2,13) in the list we strike out the P-pair (2,13).

Thus we are left with the pairs

$$(3,12)$$
$$(7,8)$$

From the top the $1^{st}(j = 1)$ pair is (3,12) but the replacement does not violates S_3. Since the replacement of the P-pair (7,8) violates S_7 we replace 15 in the partition by the P-pair (7,8). The resulting partition is

$$\pi = 14 + 11 + 10 + 8 + 7 + 4 + 1 \epsilon P'_A(n).$$

e.g.3, $\psi = 15 + 11 + 10 + 5 + 4$ Here j = 1.

We list the P-pairs (α, β) vertically one by one

$$(1,14)$$
$$(2,13)$$
$$(3,12)$$
$$(6,9)$$
$$(7,8)$$

Since there are two P-pairs (5,10) and (4,11) in ψ which are connected to the P-pair (6,9) in the list we strike out the P-pairs (6,9) and (7,8).

Thus we are left with the P-pairs

$$(1,14)$$
$$(2,13)$$
$$(3,12)$$

From the top the $1^{st}(j = 1)$ P-pair is (1,14) but the replacement does not violates S_1. Since the replacement of the P-pair (3,12) violates S_3 we replace 15 in the partition by the P-pair (3, 12).

The resulting partition is

$$\pi = 12 + 11 + 10 + 5 + 4 + 3 \epsilon P'_A(n).$$

e.g.4, $\psi = 15 + 15 + 15 + 15 + 12 + 3$ Here $j = 4$

We list the P-pairs (α, β) vertically one by one

$$(1,14)$$
$$(2,13)$$
$$(4,11)$$
$$(5,10)$$
$$(6,9)$$
$$(7,8)$$

Since there is a P-pair (3,12) in ψ which is connected to the P-pair (4,11) in the list we strike out the P-pair (4,11).

Thus we are left with the P-pairs

$$(1,14)$$
$$(2,13)$$
$$(5,10)$$
$$(6,9)$$
$$(7,8)$$

From the top the $4^{th}(j = 4)$ P-pair is (6,9) but the replacement does not violate S_6. Since the replacement of the P-pair (7,8) violates S_7 we replace four 15's in the partition by the P-pairs

$$(2, 13), (5, 10), (6, 9), (7, 8)$$

The resulting partition is

$$\pi = 13 + 12 + 10 + 9 + 8 + 7 + 6 + 5 + 3 + 2\epsilon P'_A(n).$$

This proves Lemma 2. □

For $n < (2k - a - \frac{\lambda}{2} + 1)(\lambda + 1)$,

$$\begin{aligned} B_{\lambda,k,a}(n) &= \{Q^1 \cup, \ldots, \cup Q^{a-1}\} \\ &= \text{Cardinality of } P'_A(n) \\ &= A_{\lambda,k,a}(n) \end{aligned}$$

when $n = (2k - a - \frac{\lambda}{2} + 1)(\lambda + 1) + x$, $0 \le x \le \frac{\theta(\theta-1)}{2}$

$$B_{\lambda,k,a}(n) = \{Q^1 \cup, \ldots, \cup Q^{a-1}\} \cup R(n)$$

where $R(n)$ is already defined. Since $R(n) = B_{\lambda,k,a}(x)$ proof of the Revised conjecture follows from Lemma 2.

References

[1] Andrews, G.E., A Generalization of the Gollnitz-Gordon Partition Theorems, *Proc. Amer. Math. Soc.* **18** (1967), 945–952.

[2] Andrews, G.E., A generalization of the classical partition theorems, *Trans. Amer. Math. Soc.* **145** (1969), 205–221.

[3] Andrews. G.E., Partition Identities, *Advances in Math.*, **9** (1972), 10-51.

[4] Andrews, G.E., On the general Rogers-Ramanujan theorem, *Mem. Amer. Math. Soc.*, No. 152, 1974, pp. 1–86.

[5] Andrews, G.E., Bessenrodt, C. and Olsson, J.B., Partition identities and labels for some modular characters, *Trans. Amer. Math. Soc.* **344** (1994), 597–615.

[6] Gollnitz. H., Partitionen mit Differenzenbedingngen, *J.Reine und angew. Math.* 225, (1967), 154-190.

[7] Gordon. B., A Combinatorial Generalization of the Rogers-Ramanujan Identities, *Amer. J. Math.* **83** (1961), 393-399.

[8] Padmavathamma and T.G. Sudha, On a Conjecture of Andrews, *Internat. J. Math and Math. Sci.* Vol.16, No.4 (1993), 763-774.

[9] Padmavathamma and Ruby Salestina. M., On a Conjecture of Andrews - II, *Number Theory and Discrete Mathematics*, Proceedings of the in Honour of Srinivasa Ramanujan, Chandigarh, India, 2000, 135-147.

[10] Padmavathamma and Ruby Salestina. M., 1974 Conjecture of Andrews on partitions, *Internat. J. Math and Math. Sci.* (to appear).

[11] Padmavathamma and Ruby Salestina. M., A Combinatorial proof of a Theorem of Andrews on partitions (Communicated).

[12] Padmavathamma, Ruby Salestina. M. and Sudarshan.S.R., A new Theorem on partitions, *Proceedings of the International Conference on Special Functions*, IMSC, Chennai, India, 23 - 27, September 2002 (to appear).

[13] Padmavathamma, Chandrashekara, B.M, Raghavendra.R and Christian Krattentalher, Analytic Proof of the Partition Identity $A_{5,3,3} = B^0_{5,3,3}(n)$, *Ramanujan Journal* (to appear).

[14] Schur. I.J., Zur additiven Zahlentheorie *Sitzungsber. Deutsch. Adak. Wissensch. Berlin*, Phys-Math. K1, (1926), 488-495.

Padmavathamma[1], Raghavendra[2] R and Chandrashekara[3] B.M.
Department of Studies in Mathematics,
University of Mysore, Manasagangotri,
Mysore – 570 006, Karnataka,
India.

e-mail: [1]padma_vathamma@yahoo.com,
[2]raghu_maths@yahoo.co.in
[3]chandru_alur@yahoo.com

The Riemann Zeta Function and Related Themes – 2006, pp. 57–64

Davenport constant and non-Abelian version of Erdős-Ginzburg-Ziv theorem

W.D. Gao and R. Thangadurai

Dedicated to Professor K. Ramachandra on his 70th birthday

Abstract

In this article, apart from giving a survey of known results on Davenport constant for finite groups, we shall prove the following new result. Let G be a non-abelian group with $\mathbf{Z}(G)$ as its center. Let $S = (a_1, a_2, \ldots, a_\ell)$ be a sequence in G of length $\ell = |G| + D(G) - 1$, where $D(G)$ is the Davenport constant (see below, definition 1) for the group G. Suppose that there exists $g \in \mathbf{Z}(G)$ such that g appears in S maximum number of times. Then, there exist distinct integers $i_1, i_2, \ldots, i_{|G|}$ from $1, 2, \ldots, \ell$ such that the product $a_{i_1} a_{i_2} \ldots a_{i_{|G|}}$ is the identity element in G.

1 Introduction

In 1961, Erdős, Ginzburg and Ziv [2] proved that given any sequence $a_1, a_2, \ldots, a_{2n-1}$ (not necessarily distinct) of elements in $\mathbb{Z}_n$, the cyclic group of order n, there exists a subsequence with n elements whose sum is the identity in $\mathbb{Z}_n$. Moreover, they proved that $2n-1$ cannot be replaced by $2n-2$.

This theorem is a cornerstone of many questions in 'Zero-sum Problems' which is now one of the active fields of research in Combinatorial Number Theory. The above theorem has many generalizations (See [25] for instance).

From now onwards, we denote any finite group (not necessarily abelian) by G which is additively written.

Definition 1. By $D(G)$, we denote the smallest positive integer t such that given any sequence $g_1, g_2, \ldots, g_\ell$ in G with $\ell \geq t$ there exist distinct integers $1 \leq i_1 < i_2 < \cdots < i_r \leq \ell$ and a permutation π on the symbols $\{1, 2, \ldots, r\}$ such that $g_{i_{\pi(1)}} + g_{i_{\pi(2)}} + \cdots + g_{i_{\pi(r)}} = 0$ in G.

When G is abelian, $D(G)$ is nothing but the well-known *Davenport constant*. This above generalization is considered, for example, in [3]. Clearly, we have $D(G) \leq |G|$.

In the definition 1, if we take π to be identity, then $D(G)$ is called the Strong Davenport constant and we denote it by $d(G)$.

Definition 2. By $ZS(G)$, we denote the smallest positive integer t such that given any sequence $g_1, g_2, \ldots, g_\ell$ in G with $\ell \geq t$ there exists distinct integers $1 \leq i_1 < i_2 < \cdots < i_{|G|} \leq \ell$ and a permutation π on the symbols $\{1, 2, \ldots, |G|\}$ such that $g_{i_{\pi(1)}} + g_{i_{\pi(2)}} + \cdots + g_{i_{\pi(|G|)}} = 0$ in G.

When G is abelian, the first author [7] proved that

$$ZS(G) = |G| - 1 + D(G),$$

1991 *Mathematics Subject Classification.* Primary 11B75, Secondary 20K99.
Key words and phrases. zero-sum sequences, non-abelian groups, Erdős, Ginzburg and Ziv Theorem

which had been earlier conjectured by Hamidoune. When $G = \mathbb{Z}_n$, this result is nothing but the Erdös, Ginzburg and Ziv theorem.

An interesting problem is to prove or disprove the following conjecture.

Conjecture 1. (Gao and Zhuang, [12]) For every finite group G, we have $ZS(G) = |G| - 1 + D(G)$.

J.E. Olson [21] proved that $ZS(G) \leq 2|G| - 1$. Recently, Dimitrov [4] gave a very simple proof of this fact for all solvable groups. In 1984, Peterson and Yuster [23] proved that $ZS(G) \leq 2|G| - 2$ for a non-cyclic group G. In 1988, for a positive integer r, with the restriction that $|G| \geq 600((r-1)!)^2$, Yuster [27] proved that $ZS(G) \leq 2|G| - r$; for a non-cyclic solvable group G and 1996, the first author [8] proved that $ZS(G) \leq \frac{11}{6}|G| - 1$ for a non-cyclic solvable group G. In 2003, Dimitrov [5] proved a stronger result when G is a non-cyclic p group. More precisely, he proved:

Let $s \geq \left(1 + \frac{2p-1}{p^2}\right)|G| - 1$ be any integer. Then for any sequence $S = (a_1, a_2, \ldots, a_s)$ in G, there exist $1 \leq i_1 < i_2 < \cdots < i_{|G|} \leq s$ such that $a_{i_1} + a_{i_2} + \cdots + a_{i_{|G|}} = 0$ in G.

Using the method of the proof of the above result, in [6], Dimitrov mentions that $ZS(G) \leq \frac{7}{4}|G| - 1$ for all non-cyclic nilpotent groups G. More recently, Gao and Juan [12] proved Conjecture 1 for any Dihedral G of large prime index.

In this article, we shall survey the known results and conjectures on Davenport constant for all finite groups and prove the following new result which is related to Conjecture 1.

Main Theorem. *Let G be a finite non-abelian group and $S = (a_1, a_2, \ldots, a_\ell)$ a sequence in G of length $\ell = |G| + D(G) - 1$. Suppose that there exists $g \in \mathbf{Z}(G)$, where $\mathbf{Z}(G)$ is the center of G such that g appears in S maximum number of times. Then, there exist distinct integers $i_1, i_2, \ldots, i_{|G|}$ from $1, 2, \ldots, |G| - 1 + D(G)$ such that $a_{i_1} + a_{i_2} + \cdots + a_{i_{|G|}} = 0$ in G.*

Notations. Let $S = (a_1, a_2, \ldots, a_\ell)$ be a sequence in G. We denote the length ℓ of S by $|S|$. If T is a subsequence of S, then ST^{-1} is a sequence obtained by deleting the terms of T from S. If T_1 and T_2 are two disjoint subsequence of S, then we write T_1T_2 for a subsequence of S after a permutation on T_1 and T_2. We call a sequence S of length ℓ to be a *zero-sum sequence* if there is a permutation π on the symbols $\{1, 2, \ldots, \ell\}$ such that $g_{\pi(1)} + g_{\pi(2)} + \cdots + g_{\pi(\ell)} = 0$ in G. We call a sequence in G a *zero-free sequence* if none of its subsequences is a zero-sum sequence. We call an element g of G to be a *hole* of S if $g \notin \sum(S) \cup \{0\}$, where $\sum(S)$ denotes the set of all possible finite sums of elements of S.

2 Davenport Constant for any finite group

One can easily see that $D(G) \leq |G|$ for any finite group G.

2.1 Davenport constant for abelian groups.

Since G is a finite abelian group, by the structure theorem, we have

$$G \cong \mathbb{Z}_{n_1} \oplus \mathbb{Z}_{n_2} \oplus \cdots \oplus \mathbb{Z}_{n_r}$$

where $1 < n_1 | n_2 | \cdots | n_r$ and $|G| = n_1 n_2 \ldots n_r$. Here r is called the *rank* (denoted by $r(G)$) of G and n_r is called the *exponent* (denoted by $\exp(G)$) of G. Define

$$M(G) = 1 + \sum_{i=1}^{r} (n_i - 1) \tag{1}$$

and

$$\kappa(G) = \sum_{i=1}^{r} n_i. \tag{2}$$

It is easy to see that $D(G) = |G|$, whenever G is cyclic. Olson ([21] and [22]) proved that $D(G) = M(G)$ for all p-groups G and for all G with rank 2. These are the major classes of groups for which this constant is known explicitly. The first author [9] studied this constant when G is of rank 3 and he conjectured that

Conjecture 2. (Gao, [9]) *For all groups G of rank 3, we have $D(G) = M(G)$.*

It is known from the results in [13], [11] and [16] that $D(G) > M(G)$ for infinitely many groups G.

Recently, P. Rath, K. Srilakshmi and the second author proved the following theorem in [24].

Theorem 1. *Let G be a finite abelian group of rank d and of exponent n. Let $\ell_1, \ell_2, \cdots, \ell_k$ and r be integers such that $1 \leq \ell_i \leq n-2$ for all $i = 1, 2, \cdots, k$ for some integer $k \geq 0$ and the positive integer*

$$r := \begin{cases} n + \left[n\left(\sum_{i=1}^{k} \log \ell_i - \log \dfrac{n^{k+1}}{|G|} \right) \right] & \text{if } \prod_{i=1}^{k} \ell_i > \dfrac{n^{k+1}}{|G|} \\ n & \text{otherwise} \end{cases}$$

Let

$$S = (\underbrace{g_1, \cdots, g_1}_{(n-\ell_1)\ times}, \cdots, \underbrace{g_k, \cdots, g_k}_{(n-\ell_k)\ times}, c_1, c_2, \cdots, c_r)$$

be a sequence in G of length

$$\rho = \begin{cases} \sum_{i=1}^{k}(n-\ell_i) + r & \text{if } k > 0 \\ n\left(1 + \log \frac{|G|}{n}\right) & \text{if } k = 0. \end{cases} \; .$$

Then S has a subsequence whose product is identity in G.

Corollary 1. *If S is a given sequence in G of length*

$$\ell = n\left(k + 1 + \log\left[\left(\frac{n-2}{n}\right)^k \frac{|G|}{n}\right]\right) - k,$$

where n is the exponent of G and $k \geq 0$ denotes the number of distinct elements of G that are repeated at least twice in the given sequence. Then S contains a non-empty subsequence whose product is identity.

The Corollary 1 improves the previous best upper bound due to Alford, Granville and Pomerance[1] and Meshulam [17] in 1993. Also, Corollary 1 can be improved if one proves the following conjectural bound suggested by W. Narkiewicz and J. Śliwa, [18] in 1982.

Conjecture 3. (Narkiewicz and Śliwa, [18]) *For all finite abelian group G, we have, $D(G) \leq \kappa(G)$.*

Conjecture 3 has been verified for many particular groups; for more information one can refer to [10]. In 2003, Dimitrov [4] proved that $D(G) \leq c(r)\kappa(G)$ where $c(r)$ is a positive constant which depends only on r. Recently the second author [25] improved this bound to

$$D(G) < \begin{cases} (c(r) - r - 3)\kappa(G) & \text{if } 4 \leq n_1 \leq 2^{r-1} - 1 \\ (c(r) - r\ell_r)\kappa(G) & \text{if } n_1 \geq 2^{r-1}, \end{cases}$$

where $c(r)$ is a constant depending only on r and

$$\ell_r = \frac{(2^{r-1} - 1)(r - 1) + 1}{r(r - 1)}.$$

Definition 3. We denote by $\nu(G)$ the smallest non-negative integer t such that every zero-free sequence S in G of length $t \geq \nu$ has all its holes in some proper coset of G, i.e., $G\backslash(\sum(S) \cup \{0\}) \subset a + H$ for some proper subgroup H and for some $a \in G\backslash H$.

Note that $\nu(\mathbb{Z}_2) = 0$. The first author [9] proved that $\nu(G) + 1 \leq D(G) \leq \nu(G) + 2$. Also, it is known by the result of van Emde Boas [28] that if G is either cyclic or a p-group, then $D(G) = \nu(G) + 2$. The following was conjectured by the first author;

Conjecture 4. (Gao, [9]) *For every finite abelian group G, we have $D(G) = \nu(G) + 2$.*

In 2003, Dimitrov gave an upper bound for $D(G)$ when $G = \mathbb{Z}_n^d$ using covering systems modulo n. More precisely, we define the terminology as follows;

Definition 4. Let $n > 1$ be any positive integer. Let

$$M = \begin{pmatrix} a_{11} & a_{12} & \dots & a_{1r} \\ a_{21} & a_{22} & \dots & a_{2r} \\ \dots & & \dots & \dots \\ a_{m1} & a_{m2} & \dots & a_{mr} \end{pmatrix} a_{ij} \in \mathbb{Z}_n$$

be an $m \times r$ matrix over $\mathbb{Z}_n$. Let $R = (a_1 a_2 \dots a_m)$ be a $m \times 1$ column vector over $\mathbb{Z}_n$. We call the tuple (M, R) to be a *covering system* if for every r-tuple $(x_1, x_2, \dots, x_r) \in \mathbb{Z}^r$, there exists $i \in \{1, 2, \dots, m\}$ such that

$$\sum_{j=1}^{r} a_{ij}x_j \equiv a_i \pmod{n}$$

has a solution.

A finite dimensional matrix M over $\mathbb{Z}_n$ is said to be a *cover* if it is the matrix of some covering system.

For any given positive integers r and n, we denote $c(n, r)$ to the least positive integer t such that all $t \times r$ matrices over $\mathbb{Z}_n$ are covers.

Using these notions, Dimitrov [4] proved that for all positive integers r and $n > 1$, we have,

$$D(\mathbb{Z}_n^r) \leq c(n, r).$$

and he conjectured the following.

Conjecture 5. (Dimitrov, [4]) For all positive integers r and $n > 1$, we have

$$D(\mathbb{Z}_n^r) = c(n, r).$$

2.2 Davenport constant for non-abelian groups.

Olson and White [22] proved that $D(G) \leq (|G| + 1)/2$ for any non-cyclic group G. Also, trivially one observes that $D(G) \leq d(G)$. Since every group of order p is cyclic, the first non-trivial class of non-abelian group is of order $2p$ where p is an odd prime. If G is of order $2p$, then the first author and Zhuang [12] proved that $D(G) = p + 1$. Recently, Dimitrov [4] proved that $D(G \oplus \mathbb{Z}_{|G|}) \leq 2|G| - 1$ for any finite solvable group G. If H is a normal subgroup of G, then Delorme *et al.*, [3] proved that $D(G) \geq D(H) + D(G/H) - 1$. It is obvious from the definition that $D(G) \leq d(G)$. So, any reasonable bound for $d(G)$ gives a bound for $D(G)$ as well. But, providing a reasonable bound for $d(G)$ seems to be another hard problem. We have the following conjecture of Dimitrov [6].

Conjecture 6. (Dimitrov, [6]) Let G be any finite group whose complex irreducible representations (up to equivalence) have degrees $d_1, d_2, \ldots, d_r$, then

$$d(G) \leq \sum_{i=1}^{r} d_i.$$

When G is a non-abelian p-group, then we have better known result for $d(G)$. Since G is a p-group, $\mathbb{F}_pG$ is a group algebra. Its Jacobson radical J is an augmentation ideal and is nilpotent. Then its nilpontency class is called the *Loewy length* of $\mathbb{F}_pG$ and is denoted by $L(G)$. Dimitrov [5] proved that

$$d(G) \leq L(G).$$

He conjectured the following;

Conjecture 7. (Dimitrov, [5]) For all finite p-group G, we have $d(G) = L(G)$.

Note that Conjecture 7 is true for abelian p-group by the result of Olson [21] and Jennings [14].

3 Proof of Main Theorem

We start this section with the statement of the following deep theorem of Kemperman [15].

Theorem 2. *If A and B are two non-empty finite subsets of a group G such that $0 \in A \cap B$, and $0 = a + b, a \in A, b \in B$ implies that $a = b = 0$. Then,*

$$|A + B| \geq |A| + |B| - 1.$$

Lemma 1. *Let S be a sequence in G of length at least $|G|$. Let $g \in G$ be the element appearing in S maximum number of, say h, times. Then $0 \in \sum_{\leq h}(S)$ where $\sum_{\leq h}(S) = \cup_{i=1}^{h} \sum_i(S)$ and $\sum_i(S)$ denotes the set of all possible sums of i elements of S.*

Proof: One can distribute S into h non-empty subsets $B_1, B_2, \ldots, B_h$, such that $\sum |B_i| = |S|$. For any two nonempty subsets A, B of G, let $A \oplus B = A \cup B \cup (A + B)$, and this definition can be generalized to three or more subsets by induction.

Assume to the contrary that $0 \notin \sum_{\leq h}(S)$, then $0 \notin B_i$ and

$$0 \notin B_1 \oplus B_2 \subset B_1 \oplus B_2 \oplus B_3 \subset \cdots \subset B_1 \oplus B_2 \oplus B_3 \oplus \cdots \oplus B_h.$$

Set $A_i = \{0\} \cup B_i$ for $i = 1, \dots, h$. Then, by Theorem 2, we obtain, $|A_1 + A_2| \geq |A_1| + |A_2| - 1 = |B_1| + |B_2| + 1$. Since $0 \notin B_1 \oplus B_2 \oplus B_3$, one can apply Theorem 2 to $A_1 + A_2 = \{0\} \cup (B_1 \oplus B_2)$ and $A_3 = \{0\} \cup B_3$ to get

$$|A_1 + A_2 + A_3| \geq |A_1 + A_2| + |A_3| - 1 \geq |B_1| + |B_2| + 1 + |B_3| + 1 - 1 = |B_1| + |B_2| + |B_3| + 1.$$

Continuing this process, finally we arrive at:

$$|A_1 + A_2 + \cdots + A_h| \geq |B_1| + |B_2| + \cdots + |B_h| + 1 = |G| + 1,$$

a contradiction. □

Proof of the Main Theorem: Let $S = (a_1, a_2, \dots, a_\ell)$ where $\ell = |G| + D(G) - 1$ be a sequence in G. By our hypothesis, we know that some element, say, $a_\ell \in \mathbf{Z}(G)$ is repeated maximum number of, say h, times in S. We can assume that $h \leq |G| - 1$; otherwise, we are done. As $a_\ell \in \mathbf{Z}(G)$, we have $a_\ell + x = x + a_\ell$ for every $x \in G$. Therefore, we can translate, if necessary, the given sequence by a_ℓ and can assume that 0, the zero element is repeated h times. Thus, we have,

$$S = \left(a_1, a_2, \dots, a_{\ell-h}, \underbrace{0, 0, \dots, 0}_{h \text{ times}} \right) \text{ and } S_1 = (a_1, a_2, \dots, a_{\ell-h}).$$

Clearly, $\ell - h = |G| - 1 + D(G) - h \geq D(G)$.

We distinguish two cases here.

Case (i). ($\ell - h \leq |G|$)

Since $\ell - h \geq D(G)$, by the definition of $D(G)$, there exist distinct indices $i_1, i_2, \dots, i_k$ from $1, 2, \dots, \ell - h$ such that $a_{i_1} + a_{i_2} + \cdots + a_{i_k} = 0$. Choose k to be the maximal possible integer t such that this happens in S_t. This can be done by applying all possible permutations on the indices $\{1, 2, \dots, \ell - h\}$. Hence by the maximality of k, it is clear that $|G| - 1 + D(G) - h - k \leq D(G) - 1$, in turn this implies $k \geq |G| - h$ and hence $|G| - h \leq k \leq |G|$. Therefore, we can get

$$\underbrace{0 + 0 + \cdots + 0}_{|G|-k \text{ times}} + a_{i_1} + a_{i_2} + \cdots + a_{i_k} = 0 \text{ in } G,$$

as desired.

Case (ii) ($\ell - h \geq |G| + 1$)

By Lemma 1, one can find t disjoint zero-sum subsequences $T_1, \dots, T_t$ of S_1 such that $2 \leq |T_i| \leq h$, and that $|S_1(T_1 \dots T_t)^{-1}| \leq |G| - 1$. Let W be the maximal zero-sum subsequence of $S_1(T_1 \dots T_t)^{-1}$ (if it exists). If $|W| \geq |G| - h$, then we are done, by the argument as in Case (i). Otherwise, $|W| \leq |G| - h - 1$. By the maximality of W, we see that $|S_1(T_1 \dots T_t)^{-1} W^{-1}| \leq D(G) - 1$. Therefore,

$$|W| + |T_1| + \cdots + |T_t| \geq \ell - h - (D(G) - 1) \geq |G| - h.$$

Note that since $2 \leq |T_i| \leq h$, we infer that, $|G| - h \leq |W| + |T_1| + \cdots + |T_k| \leq |G|$ for some $k \in \{1, 2, \dots, t\}$. But $WT_1 \dots T_k$ is zero-sum and we are done, as we have h number of 0's outside S_1. □

Acknowledgment. The first author is supported by NSFC with grant number 10271080. We are grateful to the referee for many useful suggestions for the better presentation of the paper. Also we are thankful to him/her for pointing out typos.

References

[1] W.R. Alford, A. Granville and C. Pomerance, There are infinitely many Carmichael numbers, *Annals of Math.*, **139** (2) (1994), no. 3, 703-722.

[2] P. Erdös, A. Ginzburg and A. Ziv, Theorem in the additive number theory, *Bull. Res. Council Israel*, **10 F**(1961), 41-43.

[3] C. Delorme, O. Ordaz and D.Quiroz, Some remarks on Davenport constant, *Discrete Math.*, **237** (2001), 119-128.

[4] V. Dimitrov, Zero-sum problems in finite groups, Research Science Institute Students Reports, 2003, *The center for excellence in education, Vienna*, pg. 9-18.

[5] V. Dimitrov, On the strong Davenport constant of non-abelian finite p-groups, *Math. Balkanica (N.S.)*, **18** (2004), no. 1-2, 131-140.

[6] V. Dimitrov, A zero-sum result for non-abelian finite groups, *Preprint*, 2004.

[7] W.D. Gao, A combinatorial problem on finite abelian group, *J. Number Theory*, **58** (1996), 100-103.

[8] W.D. Gao, An improvement of Erdős-Ginzburg-Ziv theorem, *Acta Math. Sinca*, **39** (1996), 514-523.

[9] W.D. Gao, On Davenport's constant of finite abelian groups with rank three, *Discrete Math.*, **222** (2000), no. 1-3, 111-124.

[10] W.D. Gao, On a combinatorial problem connected with factorizations. *Colloq. Math.*, **72** (1997), no. 2, 251-268.

[11] W.D. Gao and A. Geroldinger, On zero-sum sequences in $\mathbb{Z}/n\mathbb{Z} \oplus \mathbb{Z}/n\mathbb{Z}$, *Integers*, **3** (2003), A8, 45 pp. (electronic).

[12] W.D. Gao, J.J. Zhuang, Sequences not containing long zero-sum subsequences, *European J. Combin.* **27** (2006), no. 6, 777–787.

[13] A. Geroldinger and R. Schneider, On Davenport's constant, *J. Combin. Theory, Ser. A*, **61** (1992), no. 1, 147-152.

[14] S.A. Jennings, The structure of the group ring of a p-group over modular field, *Trans. Amer. Math. Soc.*, **50** (1941), 175-185.

[15] J.H. Kemperman, On complexes in a semigroup, *Nederl. Akad. Wetensch. Proc. Ser. A.*, **59** (1956), 247-254.

[16] M. Mazur, A note on the growth of Davenport's constant, *Manuscripta Math.*, **74** (1992), no. 3, 229-235.

[17] R. Meshulam, An uncertainity inequality and zero subsums, *Discrete Math.*, **84** (1990), 197-200.

[18] W. Narkiewicz and J. Śliwa, Finite abelian groups and factorization problems - II, *Colloq. Math.*, **46** (1982), 115-122.

[19] J.E. Olson, A combinatorial problem in finite abelian groups, I, *J. Number Theory*, **1** (1969), 8-10.

[20] J.E. Olson, A combinatorial problem in finite abelian groups, II, *J. Number Theory*, **1** (1969), 195-199.

[21] J.E. Olson, On a Combinatorial Problem of Erdős, Ginzburg, and Ziv, *J. Number Theory*, **8** (1976), 52-57.

[22] J.E. Olson and E. T. White, Sums from a sequence of group elements, *Number theory and algebra,*, Academic Press, New York, (1977), 215-222.

[23] B. Peterson and T. Yuster, A generalization of an addition theorem for solvable groups, *Canad. J. Math.*, **36** (1984), 529-536.

[24] P. Rath, K. Srilaskhmi and R. Thangadurai, On Davenport's Constant, *To appear in:* International. J. Number Theory.

[25] R. Thangadurai, Non-canonical extensions of Erdő-Ginzburg-Ziv Theorem, *Integers*, (2002), A7, 14pp (electronic).

[26] R. Thangadurai, A remark on an upper bound for Davenport's constant, Preprint, 2004.

[27] T. Yuster, Bounds for counter-example to addition theorem in solvable groups, *Arch. Math.*, **51** (1988), 223-231.

[28] P. van Emde Boas, A combinatorial problem on finite abelian groups, I and II, ZW-1969-007, Math. Centre, Amsterdam.

[29] P. van Emde Boas and D. Kruyswijk, A combinatorial problem on finite abelian groups, III, ZW-1969-008, Math. Centre, Amsterdam.

W. D. Gao
Center for Combinatorics,
Nankai University,
Tianjin 300071,
China.

e-mail: wdgao_1963@yahoo.com.cn

R. Thangadurai
School of Mathematics,
Harish-Chandra Research Institute,
Chhatnag Road,Jhunsi,
Allahabad - 211019, India.

e-mail: thanga@hri.res.in

The Riemann Zeta Function and Related Themes – 2006, pp. 65–80

Extreme values of $|\zeta(1+it)|$

Andrew Granville and K. Soundararajan

Dedicated to Professor K. Ramachandra on his 70th birthday

1 Introduction

Improving on a result of J.E. Littlewood, N. Levinson [3] showed that there are arbitrarily large t for which $|\zeta(1+it)| \geq e^{\gamma}\log_2 t + O(1)$. (Throughout $\zeta(s)$ is the Riemann-zeta function, and $\log_j$ denotes the j-th iterated logarithm, so that $\log_1 n = \log n$ and $\log_j n = \log(\log_{j-1} n)$ for each $j \geq 2$.) The best upper bound known is Vinogradov's $|\zeta(1+it)| \ll (\log t)^{2/3}$.

Littlewood had shown that $|\zeta(1+it)| \lesssim 2e^{\gamma}\log_2 t$ assuming the Riemann Hypothesis, in fact by showing that the value of $|\zeta(1+it)|$ could be closely approximated by its Euler product for primes up to $\log^2(2+|t|)$ under this assumption. Under the further hypothesis that the Euler product up to $\log(2+|t|)$ still serves as a good approximation, Littlewood conjectured that $\max_{|t|\leq T}|\zeta(1+it)| \sim e^{\gamma}\log_2 T$, though later he wrote in [5] (in connection with a q-analogue): "*there is perhaps no good reason for believing ... this hypothesis*".

Our Theorem 1 evaluates the frequency with which such extreme values are attained; and if this density function were to persist to the end of the viable range then this implies the conjecture that

$$\max_{t\in[T,2T]}|\zeta(1+it)| = e^{\gamma}(\log_2 T + \log_3 T + C_1 + o(1)), \tag{1}$$

for some constant C_1. In fact it may be that $C_1 = C + 1 - \log 2$, where

$$C = \int_0^2 \log I_0(t)\frac{dt}{t^2} + \int_2^{\infty}(\log I_0(t) - t)\frac{dt}{t^2} = -.3953997\ldots,$$

and $I_0(t) := \mathbb{E}(e^{\mathrm{Re}(tX)}) = \sum_{n=0}^{\infty}(t/2)^{2n}/n!^2$ is the Bessel function (with X a random variable equidistributed on the unit circle). In Theorem 2 we show that there are arbitrarily large t for which $|\zeta(1+it)| \geq e^{\gamma}(\log_2 t + \log_3 t - \log_4 t + O(1))$, which improves upon Levinson's result but falls a little short of our conjecture.

Levinson also showed that $1/|\zeta(1+it)| \geq \frac{6e^{\gamma}}{\pi^2}(\log_2 t - \log_3 t + O(1))$ for arbitrarily large t. Theorem 1 exhibits even smaller values of $|\zeta(1+it)|$ and determines their frequency. Extrapolating Theorem 1 we are also led to conjecture that

$$\max_{t\in[T,2T]} 1/|\zeta(1+it)| = \frac{6e^{\gamma}}{\pi^2}(\log_2 T + \log_3 T + C_1 + o(1));$$

but only succeed in proving that $1/|\zeta(1+it)| \geq \frac{6e^{\gamma}}{\pi^2}(\log_2 t - O(1))$ for arbitrarily large t. K. Ramachandra [6] has obtained results analogous to Levinson's in short intervals, and R. Balasubramanian, Ramachandra and A. Sankaranarayanan [1] have considered extreme values of $|\zeta(1+it)|^{e^{i\theta}}$ for any $\theta \in [0, 2\pi)$.

Le premier auteur est partiellement soutenu par une bourse de la Conseil de recherches en sciences naturelles et en génie du Canada. The second author is partially supported by the National Science Foundation.

To be more precise let us define, for $T, \tau \geq 1$,

$$\Phi_T(\tau) := \frac{1}{T}\text{meas}\{t \in [T, 2T] : \ |\zeta(1+it)| > e^{\gamma}\tau\},$$
$$\text{and } \Psi_T(\tau) := \frac{1}{T}\text{meas}\left\{t \in [T, 2T] : \ |\zeta(1+it)| < \frac{\pi^2}{6e^{\gamma}\tau}\right\}.$$

Theorem 1. *Let T be large. Uniformly in the range $1 \ll \tau \leq \log_2 T - 20$ we have*

$$\Phi_T(\tau) = \exp\left(-\frac{2e^{\tau-C-1}}{\tau}\left(1 + O\left(\frac{1}{\tau^{\frac{1}{2}}} + \left(\frac{e^{\tau}}{\log T}\right)^{\frac{1}{2}}\right)\right)\right),$$

where c is a positive constant. The same asymptotic also holds for $\Psi_T(\tau)$.

With a judicious application of the pigeonhole principle we can exhibit even larger values of $|\zeta(1+it)|$, indeed of almost the same quality as the conjectured 1.

Theorem 2. *For large T the subset of points $t \in [0, T]$ such that*

$$|\zeta(1+it)| \geq e^{\gamma}(\log_2 T + \log_3 T - \log_4 T - \log A + O(1))$$

has measure at least $T^{1-\frac{1}{A}}$, uniformly for any $A \geq 10$.

One can also establish results analogous to Theorems 1 and 2 for the distribution of values of $|L(1,\chi)|$ where χ ranges over all non-trivial characters modulo a large prime p (see section 7 for further details). In fact Theorems 1 and 2 hold almost verbatim, just changing T to p. If one also averages over p in a dyadic interval $P \leq p \leq 2P$ then one can obtain asymptotics for the distribution function in the wider range $1 \ll \tau \leq \log_2 P + \log_3 P - O(1)$ (which we expect is the full range, up to the explicit value of the "$O(1)$").

As in [2] we can compare the distribution of $\zeta(1+it)$ with that of an appropriate probabilistic model. Let $X(p)$ denote independent random variables uniformly distributed on the unit circle, for each prime p. We extend X multiplicatively to all integers n: that is set $X(n) = \prod_{p^{\alpha}\| n} X(p)^{\alpha}$. We wish to compare the distribution of values of $\zeta(1+it)$ with the distribution of values of the random Euler products $L(1, X) := \prod_p (1 - X(p)/p)^{-1}$ (these products converge with probability 1). Now define

$$\Phi(\tau) = \text{Prob}\,(|L(1,X)| \geq e^{\gamma}\tau) \ \text{ and } \ \Psi(\tau) = \text{Prob}\left(|L(1,X)| \leq \tfrac{\pi^2}{6e^{\gamma}\tau}\right).$$

By the same methods one can show that $\Phi(\tau)$ and $\Psi(\tau)$ satisfy the same asymptotic as $\Phi_T(\tau)$ as in Theorem 1, but for arbitrary τ (see the remarks immediately after the proof of Theorem 1).

2 Preliminaries

We collect here some standard facts on $\zeta(s)$ which will be used later.

Lemma 1. *Let $y \geq 2$ and $|t| \geq y + 3$ be real numbers. Let $\frac{1}{2} \leq \sigma_0 < 1$ and suppose that the rectangle $\{z : \sigma_0 < Re(z) \leq 1, |Im(z) - t| \leq y + 2\}$ is free of zeros of $\zeta(z)$. Then for any $\sigma_0 < \sigma \leq 2$ and $|\xi - t| \leq y$ we have*

$$|\log \zeta(\sigma + i\xi)| \ll \log|t| \log(e/(\sigma - \sigma_0)).$$

Further for $\sigma_0 < \sigma \le 1$ *we have*

$$\log \zeta(\sigma + it) = \sum_{n=2}^{y} \frac{\Lambda(n)}{n^{\sigma+it} \log n} + O\Big(\frac{\log |t|}{(\sigma_1 - \sigma_0)^2} y^{\sigma_1 - \sigma}\Big),$$

where we put $\sigma_1 = \min\big(\sigma_0 + \frac{1}{\log y}, \frac{\sigma+\sigma_0}{2}\big)$.

Proof: The first assertion follows from Theorem 9.6(B) of Titchmarsh [8]. In proving the second assertion we may plainly suppose that $y \in \mathbb{Z} + \frac{1}{2}$. Then Perron's formula gives, with $c = 1 - \sigma + \frac{1}{\log y}$,

$$\begin{aligned}\frac{1}{2\pi i} \int_{c-iy}^{c+iy} \log \zeta(\sigma + it + w) \frac{y^w}{w} dw &= \sum_{n=2}^{y} \frac{\Lambda(n)}{n^{\sigma+it} \log n} + O\Big(\frac{1}{y} \sum_{n=1}^{\infty} \frac{y^c}{n^{\sigma+c}} \frac{1}{|\log(y/n)|}\Big) \\ &= \sum_{n=2}^{y} \frac{\Lambda(n)}{n^{\sigma+it} \log n} + O(y^{-\sigma} \log y). \qquad (3)\end{aligned}$$

We now move the line of integration to the line $\mathrm{Re}(w) = \sigma_1 - \sigma < 0$. Our hypothesis ensures that the integrand is regular over the region where the line is moved, except for a simple pole at $w = 0$ which leaves the residue $\log \zeta(\sigma + it)$. Thus the left side of (3) equals $\log \zeta(\sigma + it)$ plus

$$\frac{1}{2\pi i} \Big(\int_{c-iy}^{\sigma_1-\sigma-iy} + \int_{\sigma_1-\sigma-iy}^{\sigma_1-\sigma+iy} + \int_{\sigma_1-\sigma+iy}^{c+iy} \Big) \log \zeta(\sigma + it + w) \frac{y^w}{w} dw \ll \frac{\log |t|}{(\sigma_1 - \sigma_0)^2} y^{\sigma_1-\sigma},$$

upon using the first part of the Lemma. □

Using Lemma 1 we shall show that most of the time we may approximate $\zeta(s)$ by a short Euler product.

Lemma 2. *Let* $\frac{1}{2} < \sigma \le 1$ *be fixed and let* T *be large. Let* $T/2 \ge y \ge 3$ *be a real number. The asymptotic*

$$\log \zeta(\sigma + it) = \sum_{n=2}^{y} \frac{\Lambda(n)}{n^{\sigma+it} \log n} + O\big(y^{(\frac{1}{2}-\sigma)/2} \log^3 T\big)$$

holds for all $t \in (T, 2T)$ *except for a set of measure* $\ll T^{5/4-\sigma/2} y (\log T)^5$.

Proof: This follows upon using the zero-density result $N(\sigma_0, T) \ll T^{3/2-\sigma_0} (\log T)^5$ (see Theorem 9.19 A of [8]) and appealing to Lemma 1 (taking $\sigma_0 = (1/2 + \sigma)/2$ there). □

3 Approximating $\zeta(1 + it)$ by a short Euler product

Lemma 3. *Suppose* $2 \le y \le z$ *are real numbers. Then for arbitrary complex numbers* $x(p)$ *we have*

$$\frac{1}{T} \int_T^{2T} \Big| \sum_{y \le p \le z} \frac{x(p)}{p^{it}} \Big|^{2k} dt \ll \Big(k \sum_{y \le p \le z} |x(p)|^2 \Big)^k + T^{-\frac{2}{3}} \Big(\sum_{y \le p \le z} |x(p)| \Big)^{2k}$$

for all integers $1 \le k \le \log T / (3 \log z)$.

Proof: The quantity we seek to estimate is

$$\sum_{\substack{p_1,\dots,p_k\\ y\le p_j\le z}} \sum_{\substack{q_1,\dots,q_k\\ y\le q_j\le z}} \overline{x(p_1)\cdots x(p_k)} x(q_1)\cdots x(q_k) \frac{1}{T}\int_T^{2T} \left(\frac{p_1\cdots p_k}{q_1\cdots q_k}\right)^{it} dt.$$

The diagonal terms $p_1\cdots p_k = q_1\cdots q_k$ contribute

$$\ll k!\left(\sum_{y\le p\le z} |x(p)|^2\right)^k.$$

If $p_1\cdots p_k \neq q_1\cdots q_k$ then as both quantities are below $z^k \le T^{\frac{1}{3}}$ we have that

$$\frac{1}{T}\int_T^{2T}\left(\frac{p_1\cdots p_k}{q_1\cdots q_k}\right)^{it} dt \ll \frac{1}{T|\log(p_1\cdots p_k/q_1\cdots q_k)|} \ll T^{-\frac{2}{3}}.$$

Hence the off diagonal terms contribute $\ll T^{-\frac{2}{3}}\left(\sum_{y\le p\le z}|x(p)|\right)^{2k}$, proving the Lemma. □

Define $\zeta(s;y) := \prod_{p\le y}(1-p^{-s})^{-1}$.

Proposition 1. *Let T be large and let $\log T(\log_2 T)^4 \ge y \ge e^2 \log T$ be a real number. Then*

$$\zeta(1+it) = \zeta(1+it;y)\left(1 + O\left(\frac{\sqrt{\log T}}{\sqrt{y}\log_2 T}\right)\right)$$

for all $t \in (T, 2T)$ except for a set of measure at most $T\exp(-\log T/50\log_2 T)$.

Proof: Setting $z = (\log T)^{100}$ we deduce from Lemma 2 that $\zeta(1+it) = \zeta(1+it;z)(1+O(1/\log T))$ for all $t \in (T,2T)$ except for a set of measure at most $T^{4/5}$. Using Lemma 3 with $k = [\log T/(300\log_2 T)]$ and $x(p) = 1/p$ we get that

$$\begin{aligned}\frac{1}{T}\int_T^{2T}\left|\sum_{y\le p\le z}\frac{1}{p^{1+it}}\right|^{2k} dt &\ll \left(k\sum_{y\le p\le z}\frac{1}{p^2}\right)^k + T^{-\frac{2}{3}}\left(\sum_{y\le p\le z}\frac{1}{p}\right)^{2k}\\ &\ll \left(\frac{\log T}{y}\right)^k\left(\frac{1}{10\log y}\right)^{2k} + T^{-\frac{1}{3}},\end{aligned}$$

and so

$$\left|\sum_{y\le p\le z}\frac{1}{p^{1+it}}\right| \le \frac{\sqrt{\log T}}{\sqrt{y}\log y}$$

for all $t \in [T, 2T]$ except for a set of measure $\le T\exp(-\log T/49\log_2 T)$. The Proposition thus follows, by combining the above estimates, since

$$\zeta(1+it;y) = \zeta(1+it;z)\exp\left(-\sum_{y\le p\le z}\left(\frac{1}{p^{1+it}} + O\left(\frac{1}{p^2}\right)\right)\right).$$

□

4 Moments of short Euler products

In this section we show how to evaluate large moments of the short Euler products obtained in §3. Below, for any complex number z, $d_z(n)$ will denote the z-th divisor function. That is, $d_z(n)$ is the n-th Dirichlet series coefficient of $\zeta(s)^z$.

Theorem 3. *Let* $\log T(\log_2 T)^4 \geq y \geq e^2 \log T$ *be a real number. Let* $z = \delta k$ *where* $\delta = \pm 1$ *and* $2 \leq k \leq \log T/(e^{10}\log(y/\log T))$ *is an integer. Then*

$$\frac{1}{T}\int_T^{2T} |\zeta(1+it;y)|^{2z} dt = \sum_{\substack{n=1 \\ p|n \implies p\leq y}}^{\infty} \frac{d_z(n)^2}{n^2}\left(1 + O\left(\exp\left(-\frac{\log T}{2(\log_2 T)^4}\right)\right)\right)$$

$$= \prod_{p\leq k}\left(1-\frac{\delta}{p}\right)^{-2k\delta} \exp\left(\frac{2k}{\log k}\left(C + O\left(\frac{k}{y} + \frac{1}{\log k}\right)\right)\right).$$

Throughout this section let z, y, k, δ be as in Theorem 3. If $k \leq 10^6$ then we divide $[1, y]$ into the intervals $I_0 = [k, y]$ and $I_1 = [1, k)$ and take here $J := 1$. If $k > 10^6$ then we define $J := [4\log_2 k/\log 2] + 1$ and divide $[1, y]$ into the $J+1$-intervals $I_0 = [k, y]$, $I_j = [k/2^j, k/2^{j-1})$ for $1 \leq j \leq J-1$, and $I_J = [1, k/2^J) \subset [1, k/(\log k)^4]$. Given a subset R of the index set $\{0, 1, \ldots, J\}$ we define $\mathcal{S}(R)$ to be the set of integers n whose prime factors all lie in $\cup_{r\in R} I_r$. We also define

$$\zeta(s;R) := \prod_{p\in\cup_{r\in R}I_r}\left(1-\frac{1}{p^s}\right)^{-1} = \sum_{n\in\mathcal{S}(R)}\frac{1}{n^s}.$$

Proposition 2. *Let* R *be any subset of* $\{0, \ldots, J\}$. *Then we have that*

$$\frac{1}{T}\int_T^{2T} |\zeta(1+it;R)|^{2z} dt = \sum_{n\in\mathcal{S}(R)} \frac{d_z(n)^2}{n^2}\left(1 + O\left(\exp\left(-\frac{\log T}{2(\log_2 T)^4}\right)\right)\right).$$

Note that the first part of Theorem 3 follows from the case $R = \{0, 1, \ldots, J\}$. While this is the case of interest for us, the formulation of Proposition 2 is convenient for our proof which is based on induction on the cardinality of R.

Lemma 4. *For any prime* p *we have*

$$\sum_{a=0}^{\infty}\frac{d_z(p^a)^2}{p^{2a}} = I_0\left(\frac{2k}{p}\right)\exp(O(k/p^2)),$$

where I_0 *denotes the I-Bessel function. Also*

$$\left(1-\frac{\delta}{p}\right)^{-2k\delta} \geq \sum_{a=0}^{\infty}\frac{d_z(p^a)^2}{p^{2a}} \geq \frac{1}{50}\min\left(1,\frac{p}{k}\right)\left(1-\frac{\delta}{p}\right)^{-2k\delta},$$

so that if $\mathcal{P}$ *is any subset of the primes* $\leq y$ *then, uniformly,*

$$\sum_{\substack{n\geq 1 \\ p|n \implies p\in\mathcal{P}}}\frac{d_z(n)^2}{n^2} \geq T^{O(1/\log_2 T)}\prod_{p\in\mathcal{P}}\left(1-\frac{\delta}{p}\right)^{-2k\delta}.$$

Proof: Since

$$\sum_{a=0}^{\infty} \frac{d_z(p^a)^2}{p^{2a}} = \int_0^1 \left|1 - \frac{e(\theta)}{p}\right|^{-2z} d\theta = \int_0^1 \exp(O(k/p^2)) \exp\left(2\frac{z}{p}\cos(2\pi\theta)\right) d\theta$$

we obtain the first assertion. The upper bound in the second statement follows since $|1 - e(\theta)/p|^{-\delta} \le (1-\delta/p)^{-\delta}$. When $p > k$ we have that $(1-\delta/p)^{-2k\delta} \le (1-1/\max(2,k))^{-2k} \le 16$ and so the lower bound follows in this case. When $p \le k$ consider only θ such that $e(\theta)$ lies on the arc $(\delta e^{-ip/(10k)}, \delta e^{ip/(10k)})$. For such θ we may check that $|1 - e(\theta)/p|^{-2k\delta} \ge (1 - \delta/p)^{-2k\delta}(1-1/(25k))^k \ge \frac{4}{5}(1-\delta/p)^{-2k\delta}$ from which the lower bound in this case follows.

Now

$$\prod_{\substack{k<p\le y\\ p\in\mathcal{P}}} \left(1 - \frac{\delta}{p}\right)^{-2k\delta} \le \exp\left(O\left(\sum_{k<p\le y} \frac{k}{p}\right)\right) \ll \left(\frac{\log y}{\log k}\right)^{O(k)} \ll T^{O(1/\log_2 T)},$$

and

$$\sum_{\substack{n\ge1\\ p|n \implies p\in\mathcal{P}}} \frac{d_z(n)^2}{n^2} > \sum_{\substack{n\ge1\\ p|n \implies p\le k \text{ and } p\in\mathcal{P}}} \frac{d_z(n)^2}{n^2} \ge \prod_{\substack{p\le k\\ p\in\mathcal{P}}} \frac{p}{50k}\left(1 - \frac{\delta}{p}\right)^{-2k\delta},$$

which together imply the third assertion by the prime number theorem. □

Lemma 5. *Suppose* $0 \le r \le J$ *and put* $M_0 := T^{\frac{1}{5}}$ *and* $M_r = T^{\frac{1}{5r^2}}$ *for* $r \ge 1$. *Then we have that*

$$\sum_{\substack{m\in S(\{r\})\\ m\ge M_r}} \frac{2^{\omega(m)}}{m} \sum_{\ell\in S(\{r\})} \frac{|d_z(m\ell)d_z(\ell)|}{\ell^2} \le \left(\sum_{\ell\in S(\{r\})} \frac{d_z(\ell)^2}{\ell^2}\right) \exp\left(-\frac{\log T}{(\log_2 T)^4}\right).$$

Proof: Denote the left side of the estimate in Lemma 5 by N_r and let

$$D_r = \sum_{\ell\in C(\{r\})} \frac{d_z(\ell)^2}{\ell^2}.$$

For any $1 \ge \alpha > 0$ we have

$$\begin{aligned} N_r &\le M_r^{-\alpha} \sum_{m\in C(\{r\})} \frac{2^{\omega(m)}}{m^{1-\alpha}} \sum_{\ell\in C(\{r\})} \frac{|d_z(m\ell)d_z(\ell)|}{\ell^2} \\ &= M_r^{-\alpha} \prod_{p\in I_r} \left(\sum_{a=0}^{\infty} \frac{|d_z(p^a)|^2}{p^{2a}} + 2\sum_{u=1}^{\infty} \frac{1}{p^{u(1-\alpha)}} \sum_{a=0}^{\infty} \frac{|d_z(p^a)d_z(p^{u+a})|}{p^{2a}}\right). \end{aligned} \tag{4}$$

We record two bounds for the pth term of the product in (4): Firstly

$$\begin{aligned} \sum_{a=0}^{\infty} \frac{|d_z(p^a)|^2}{p^{2a}} + 2\sum_{u=1}^{\infty} \frac{1}{p^{u(1-\alpha)}} \sum_{a=0}^{\infty} \frac{|d_z(p^a)d_z(p^{u+a})|}{p^{2a}} &\le 2\sum_{a=0}^{\infty} \frac{|d_z(p^a)|}{p^{a(1+\alpha)}} \sum_{u=-a}^{\infty} \frac{|d_z(p^{u+a})|}{p^{(u+a)(1-\alpha)}} \\ &= 2\left(1 - \frac{\delta}{p^{1-\alpha}}\right)^{-\delta k}\left(1 - \frac{\delta}{p^{1+\alpha}}\right)^{-\delta k}. \end{aligned} \tag{5}$$

Secondly, since $|d_z(p^{u+a})| \le |d_z(p^a)||d_z(p^u)|$,

$$\sum_{a=0}^{\infty} \frac{|d_z(p^a)|^2}{p^{2a}} + 2\sum_{u=1}^{\infty} \frac{1}{p^{u(1-\alpha)}} \sum_{a=0}^{\infty} \frac{|d_z(p^a)d_z(p^{u+a})|}{p^{2a}}$$
$$\le \sum_{a=0}^{\infty} \frac{|d_z(p^a)|^2}{p^{2a}} \left(1 + 2\sum_{u=1}^{\infty} \frac{|d_z(p^u)|}{p^{u(1-\alpha)}}\right)$$
$$\le \sum_{a=0}^{\infty} \frac{|d_z(p^a)|^2}{p^{2a}} \left(2\left(1 - \frac{\delta}{p^{1-\alpha}}\right)^{-\delta k} - 1\right). \tag{6}$$

Now consider the case $r = 0$ and note that $k \le p$ for all $p \in I_0$. Here we use the bound (6) in (4). We choose $\alpha = 1/(10 \log_2 T)$ and note that for $p \in I_0$, $2(1 - \delta/p^{1-\alpha})^{-\delta k} - 1 \le 2(1 - e^{1/9}/p)^{-k} - 1 \le e^{4k/p}$. Hence we get that

$$N_0 \le D_0 \exp\left(-\frac{\log M_0}{10 \log_2 T} + 4k \sum_{k \le p \le y} \frac{1}{p}\right)$$
$$\le D_0 \exp\left(-\frac{\log M_0}{10 \log_2 T} + \frac{4k}{\log k} \sum_{k \le p \le y} \frac{\log p}{p}\right).$$

Now $\sum_{k \le p \le y} \log p/p \le \log(25y/k)$ (see Theorem I.1.7 of Tenenbaum [7]) and recall that $k \le \log T/(e^{10} \log(y/\log T))$ and that $M_0 = T^{1/5}$. The bound in the lemma then follows in this case.

Suppose now that $r \ge 1$ so that $p \le k$ for all $p \in I_r$. Here we use the bound (5) in (4). We take $\alpha = 1/(10 \cdot 2^{r/2} \log(ek))$ and note that for $p \le k$,

$$\left(1 - \frac{\delta}{p^{1-\alpha}}\right)^{-\delta} \left(1 - \frac{\delta}{p^{1+\alpha}}\right)^{-\delta} \left(1 - \frac{\delta}{p}\right)^{2\delta} \le \left(1 - \frac{p(p^\alpha + p^{-\alpha} - 2)}{(p-1)^2}\right)^{-1}$$
$$\le \exp\left(\frac{\log^2 p}{10 \cdot 2^r p \log^2(ek)}\right).$$

Using also the lower bound in Lemma 4 we obtain that

$$N_r \le D_r \exp\left(-\frac{\log M_r}{10 \cdot 2^{r/2} \log(ek)} + \sum_{p \in I_r} \left(\log \frac{100k}{p} + \frac{k \log p}{10 \cdot 2^r p \log(ek)}\right)\right). \tag{7}$$

If $1 \le r \le J - 1$ then we deduce that

$$N_r \le D_r \exp\left(-\frac{\log M_r}{10 \cdot 2^{r/2} \log(ek)} + \sum_{k/2^r \le p \le k/2^{r-1}} (r+5)\right)$$
$$\le D_r \exp\left(-\frac{\log M_r}{10 \cdot 2^{r/2} \log(ek)} + \frac{8(r+5)k}{2^r \log(ek)}\right)$$

and since $\log M_r = (\log T)/(5r^2)$ this gives $N_r \le D_r \exp(-\log T/(\log_2 T)^4)$ for large T. If $r = J$ and $k \le 10^6$ then the Lemma follows at once from (7). If $r = J$ and $k > 10^6$ then (7) gives that

$$N_r \le D_r \exp\left(-\frac{\log M_J}{10\cdot 2^{J/2}\log(ek)} + \sum_{p\le k/(\log k)^4}\left(\log\frac{100k}{p} + \frac{k\log p}{10\cdot 2^J p\log(ek)}\right)\right)$$
$$\le D_r \exp\left(-\frac{\log M_J}{10\cdot 2^{J/2}\log(ek)} + O\left(\frac{\log T}{(\log_2 T)^4}\right)\right),$$

which proves the Lemma in this case. □

Proof of Proposition 2 : We prove Proposition 2 by induction on the cardinality of R. The case when $R=\emptyset$ is clear and suppose the Proposition holds for all proper subsets of R. We expand

$$|\zeta(1+it;R)|^{2z} = \sum_{\substack{m_r,n_r\in S(\{r\})\\ \text{for all } r\in R}} \prod_{r\in R}\left(\frac{d_z(m_r)d_z(n_r)}{m_r n_r}\right)\left(\frac{\prod_{r\in R} m_r}{\prod_{r\in R} n_r}\right)^{it}.$$

Set $u_r = m_r n_r/(m_r,n_r)^2$. Using inclusion-exclusion we decompose the above as

$$\sum_{\substack{bm_r,n_r\in S(\{r\}),\text{ and}\\ u_r\le M_r \text{ for all } r\in R}} + \sum_{\substack{W\subset R\\ W\ne\emptyset}}(-1)^{|W|-1}\sum_{\substack{m_r,n_r\in S(\{r\})\\ \text{for all } r\in R,\text{ and}\\ u_w>M_w \text{ for all } w\in W}} \tag{8}$$

with M_w as in Lemma 5.

First let us consider the contribution of the first sum in (8). This gives

$$\sum_{\substack{m_r,n_r\in S(\{r\}),\text{ and}\\ u_r\le M_r \text{ for all } r\in R}} \prod_{r\in R}\left(\frac{d_z(m_r)d_z(n_r)}{m_r n_r}\right)\frac{1}{T}\int_T^{2T}\left(\prod_{r\in R}\frac{m_r}{n_r}\right)^{it} dt. \tag{9}$$

If we reduce $\prod_{r\in R} m_r/n_r$ to lowest terms then both the numerator and denominator would be bounded by $\prod_r u_r \le \prod_{r\in R} M_r \le T^{\frac{(1+\pi^2/6)}{5}} \le T^{\frac{3}{5}}$. Thus if $\prod_{r\in R} m_r/n_r \ne 1$ then

$$\frac{1}{T}\int_T^{2T}\left(\frac{\prod_{r\in R} m_r}{\prod_{r\in R} n_r}\right)^{it} dt \ll \frac{1}{T|\log\prod_r m_r/n_r|} \ll T^{-\frac{2}{5}}.$$

Hence we obtain that the expression in (9) equals

$$\sum_{\substack{m_r=n_r\in S(\{r\})\\ \text{for all } r\in R}} \prod_{r\in R}\left(\frac{d_z(m_r)}{m_r}\right)^2 + O\left(T^{-\frac{2}{5}} \sum_{\substack{m_r,n_r\in S(\{r\})\\ \text{for all } r\in R}} \prod_{r\in R}\left(\frac{|d_z(m_r)d_z(n_r)|}{m_r n_r}\right)\right).$$

The main term above is $\sum_{n\in S(R)} d_z(n)^2/n^2$. The error term is $\ll T^{-\frac{2}{5}}\prod_{p\in\cup_{r\in R} I_r}(1-\delta/p)^{-2k\delta}$ and using the lower bound of Lemma 4 this is $\ll T^{-\frac{1}{3}}\sum_{n\in S(R)} d_z(n)^2/n^2$. Thus the contribution of the first term in (8) is

$$(1+O(T^{-\frac{1}{3}}))\sum_{n\in S(R)}\frac{d_z(n)^2}{n^2}. \tag{10}$$

Now we consider the contribution of the second term in (8). This gives

$$\sum_{\substack{W\subset R\\ W\neq\emptyset}}(-1)^{|W|-1}\sum_{\substack{m_w,n_w\in\mathcal{S}(\{w\}),\text{ and }\\ u_w>M_w\text{ for all } w\in W}}\prod_{w\in W}\left(\frac{d_z(m_w)d_z(n_w)}{m_wn_w}\right)$$
$$\times\frac{1}{T}\int_T^{2T}\left(\frac{\prod_{w\in W}m_w}{\prod_{w\in W}n_w}\right)^{it}|\zeta(1+it;R-W)|^{2z}dt,$$

which is bounded in magnitude by

$$\sum_{\substack{W\subset R\\ W\neq\emptyset}}\sum_{\substack{m_w,n_w\in\mathcal{S}(\{w\}),\text{ and }\\ u_w>M_w\text{ for all } w\in W}}\prod_{w\in W}\left(\frac{|d_z(m_w)d_z(n_w)|}{m_wn_w}\right)\frac{1}{T}\int_T^{2T}|\zeta(1+it;R-W)|^{2z}dt.$$

By the induction hypothesis we see that

$$\frac{1}{T}\int_T^{2T}|\zeta(1+it;R-W)|^{2z}dt\ll\sum_{n\in\mathcal{S}(R-W)}\frac{d_z(n)^2}{n^2},$$

while from Lemma 5 (with $m=u_w$ and $\ell=(m_w,n_w)$ so that $d_z(m\ell)d_z(\ell)=d_z(m_w)d_z(n_w)$; and note that the number of pairs m_w,n_w which give rise to a given pair ℓ,m is exactly $2^{\omega(m)}$) we deduce that

$$\sum_{\substack{m_w,n_w\in\mathcal{S}(\{w\})\\ u_w>M_w}}\frac{|d_z(m_w)d_z(n_w)|}{m_wn_w}\leq\sum_{n\in\mathcal{S}(\{w\})}\frac{d_z(n)^2}{n^2}\exp\left(-\frac{\log T}{(\log_2 T)^4}\right).$$

From these estimates it follows that the contribution of the second term in (8) is

$$\ll|R|\sum_{n\in\mathcal{S}(R)}\frac{d_z(n)^2}{n^2}\exp\left(-\frac{\log T}{(\log_2 T)^4}\right).$$

Combining this with (10) we obtain Proposition 2. □

Proof of Theorem 3: In view of Proposition 2 it remains only to prove that

$$\sum_{\substack{n=1\\ p|n\implies p\leq y}}^{\infty}\frac{d_z(n)^2}{n^2}=\prod_{p\leq k}\left(1-\frac{\delta}{p}\right)^{-2k\delta}\exp\left(\frac{2k}{\log k}\left(C+O\left(\frac{k}{y}+\frac{1}{\log k}\right)\right)\right).\tag{11}$$

Using the first part of Lemma 4 for $p\geq\sqrt{k}$ and the second part for $p<\sqrt{k}$ we see that

$$\sum_{\substack{n=1\\ p|n\implies p\leq y}}^{\infty}\frac{d_z(n)^2}{n^2}=\prod_{p<\sqrt{k}}\left(1-\frac{\delta}{p}\right)^{-2k\delta}\prod_{\sqrt{k}\leq p\leq y}I_0\left(\frac{2k}{p}\right)\exp(O(\sqrt{k})).$$

Since $\log I_0(t)=O(t^2)$ for $0\leq t\leq 2$ we have by the prime number theorem and partial summation that

$$\sum_{k\le p\le y} \log I_0\left(\frac{2k}{p}\right) = \frac{2k}{\log k}\int_{2k/y}^{2} \log I_0(t)\frac{dt}{t^2} + O\left(\frac{k}{\log^2 k}\right)$$
$$= \frac{2k}{\log k}\int_{0}^{2} \log I_0(t)\frac{dt}{t^2} + O\left(\frac{k^2}{y\log k} + \frac{k}{\log^2 k}\right).$$

Since $\log I_0(t) = t + O(\log t)$ for $t \ge 2$ we obtain by the prime number theorem and partial summation that

$$\sum_{\sqrt{k}\le p\le k}\left(\log I_0\left(\frac{2k}{p}\right) + 2k\delta\log\left(1-\frac{\delta}{p}\right)\right) = \frac{2k}{\log k}\int_{2}^{\infty}(\log I_0(t)-t)\frac{dt}{t^2} + O\left(\frac{k}{\log^2 k}\right).$$

These estimates prove (11) and so Theorem 3 follows. □

5 Proof of Theorem 1

Let $\log T(\log_2 T)^4 \ge y \ge e^2\log T$, and let $T\Phi_T(\tau; y)$ denote the measure of points $t \in [T, 2T]$ for which $|\zeta(1+it;y)| \ge e^{\gamma}\tau$. Taking $z = k$ for an integer $3 \le k \le \log T/(e^{10}\log(y/\log T))$ in Theorem 3 and using Mertens' theorem $\prod_{p\le k}(1-1/p)^{-1} = e^{\gamma}\log k + O(1/\log^2 k)$ we get that

$$2k\int_0^{\infty}\Phi_T(t;y)t^{2k-1}dt = \frac{1}{T}\int_T^{2T} e^{-2k\gamma}|\zeta(1+it;y)|^{2k}dt$$
$$= (\log k)^{2k}\exp\left(\frac{2k}{\log k}\left(C + O\left(\frac{k}{y}+\frac{1}{\log k}\right)\right)\right). \quad (12)$$

Now

$$\int_0^{\infty}\Phi_T(t;y)dt = e^{-\gamma}(1/T)\int_T^{2T}|\zeta(1+it;y)|dt$$
$$\le e^{-\gamma}((1/T)\int_T^{2T}|\zeta(1+it;y)|^4dt)^{1/4} \ll 1$$

by Theorem 3; so, by Hölder's inequality,

$$\int_0^{\infty}\Phi_T(t;y)t^a dt \le \left(\int_0^{\infty}\Phi_T(t;y)dt\right)^{1-a/b}\left(\int_0^{\infty}\Phi_T(t;y)t^b dt\right)^{a/b}$$
$$\ll \left(\int_0^{\infty}\Phi_T(t;y)t^b dt\right)^{a/b}$$

for $a < b$. While (12) at present holds only for integer values of k, we may interpolate to non-integer value $\kappa \in (k-1,k)$ by taking $a = 2k-3,\ b = 2\kappa-1$ and then $a = 2\kappa-1,\ b = 2k-1$ in the last inequality to obtain

$$\left(\int_0^\infty \Phi_T(t;y)t^{2k-3}dt\right)^{\frac{2\kappa-1}{2k-3}} \ll \int_0^\infty \Phi_T(t;y)t^{2\kappa-1}dt \ll \left(\int_0^\infty \Phi_T(t;y)t^{2k-1}dt\right)^{\frac{2\kappa-1}{2k-1}},$$

and so we get (12) for κ by substituting (12) for $k-1$ and k into this equation.

Suppose $1 \ll \tau \le \log_2 T - 20 - \log_2(y/\log T)$ and select $\kappa = \kappa_\tau$ such that $\log\kappa = \tau - 1 - C$. Let $\epsilon > 0$ be a bounded parameter to be fixed shortly and put $K = \kappa e^\epsilon$. Observe that

$$\begin{aligned}2\kappa \int_{\tau+\epsilon}^\infty \Phi_T(t;y)t^{2\kappa-1}dt &\le 2\kappa(\tau+\epsilon)^{2\kappa-2K} \int_{\tau+\epsilon}^\infty \Phi_T(t;y)t^{2K-1}dt \\ &\le (\tau+\epsilon)^{2\kappa(1-e^\epsilon)}\left(2K\int_0^\infty \Phi_T(t;y)t^{2K-1}dt\right).\end{aligned}$$

Using (12) we observe that

$$\begin{aligned}2K\int_0^\infty \Phi_T(t;y)t^{2K-1}dt &= \left((\log\kappa+\epsilon)\exp\left(\frac{C}{\log\kappa}\left(1+O\left(\frac{1}{\log\kappa}+\frac{\kappa}{y}\right)\right)\right)\right)^{2K} \\ &= \exp\left(\frac{2\kappa(\epsilon e^\epsilon + C(e^\epsilon-1))}{\log\kappa} + O\left(\frac{\kappa}{\log^2\kappa}+\frac{\kappa^2}{y\log\kappa}\right)\right) \\ &\quad \times (\log\kappa)^{2\kappa(e^\epsilon-1)}\int_0^\infty \Phi_T(t;y)t^{2\kappa-1}dt.\end{aligned}$$

We conclude from the last two displayed equations

$$\begin{aligned}2\kappa\int_{\tau+\epsilon}^\infty \Phi_T(t;y)t^{2\kappa-1}dt &= \exp\left(\frac{2\kappa}{\log\kappa}(1+\epsilon-e^\epsilon) + O\left(\frac{\kappa}{\log^2\kappa}+\frac{\kappa^2}{y\log\kappa}\right)\right) \\ &\quad \times \int_0^\infty \Phi_T(t;y)t^{2\kappa-1}dt.\end{aligned}$$

Choose $\epsilon = c(1/\tau + (\log T)/y)^{\frac{1}{2}}$ for a suitable constant $c > 0$, so that for large τ (and hence large κ),

$$\int_{\tau+\epsilon}^\infty \Phi_T(t;y)t^{2\kappa-1}dt \le \frac{1}{100}\int_0^\infty \Phi_T(t;y)t^{2\kappa-1}dt,$$

say. A similar argument reveals that

$$\int_0^{\tau-\epsilon} \Phi_T(t;y)t^{2\kappa-1}dt \le \frac{1}{100}\int_0^\infty \Phi_T(t;y)t^{2\kappa-1}dt.$$

Combining these two assertions with (12) for κ we obtain

$$\int_{\tau-\epsilon}^{\tau+\epsilon} \Phi_T(t;y)t^{2\kappa-1}dt = (\log\kappa)^{2\kappa}\exp\left(\frac{2\kappa C}{\log\kappa}(1+O(\epsilon^2))\right).$$

Since Φ_T is a non-increasing function we deduce that the left side above is

$$\ge \Phi_T(\tau+\epsilon;y)\tau^{2\kappa}\exp(O(\kappa\epsilon/\tau)), \qquad \text{and} \qquad \le \Phi_T(\tau-\epsilon;y)\tau^{2\kappa}\exp(O(\kappa\epsilon/\tau)).$$

It follows that

$$\Phi_T(\tau+\epsilon;y) \le \exp\left(-(2+O(\epsilon))\frac{e^{\tau-1-C}}{\tau}\right) \le \Phi_T(\tau-\epsilon;y),$$

and hence that uniformly in $\tau \le \log_2 T - 20 - \log_2(y/\log T)$ we have

$$\Phi_T(\tau;y) = \exp\left(-\frac{2e^{\tau-1-C}}{\tau}(1+O(\epsilon))\right). \tag{13}$$

From Proposition 1 we know that

$$\Phi_T(\tau) = \Phi_T(\tau+O(\epsilon);y) + O(\exp(-\log T/50\log_2 T))$$

for $\tau \ll \log_2 T$; and so from (13) we deduce that uniformly in $\tau \le \log_2 T - 20 - \log(y/\log T)$ we have

$$\Phi_T(\tau) = \exp\left(-\frac{2e^{\tau-1-C}}{\tau}(1+O(\epsilon))\right) + O\left(\exp\left(-\frac{\log T}{50\log_2 T}\right)\right).$$

Taking $y = \min(\tau\log T, (\log^2 T)/e^{10+\tau})$ above we easily obtain Theorem 1 for Φ_T. The argument for Ψ_T is analogous, using $z = -k$ in Theorem 3. □

One finds, using the first part of Lemma 4 and the observation that $\log I_0(2k/p) \ll k^2/p^2$ for $p > k$, that

$$\begin{aligned}\mathbb{E}(|L(1,X)|^{2z}) &= \sum_{n\ge 1}\frac{d_z(n)^2}{n^2} = \sum_{\substack{n=1\\ p|n \implies p\le y}}^{\infty}\frac{d_z(n)^2}{n^2}\exp\left(O\left(\frac{k^2}{y\log y}\right)\right)\\ &= \prod_{p\le k}\left(1-\frac{\delta}{p}\right)^{-2k\delta}\exp\left(\frac{2k}{\log k}\left(C+O\left(\frac{k}{y}+\frac{1}{\log k}\right)\right)\right),\end{aligned}$$

the last line following as in the proof of Theorem 3. With this estimate we can proceed precisely as in the proof of Theorem 1 to obtain the analagous estimate.

6 Large values of $|\zeta(1+it)|$: Proof of Theorem 2

Let T be large and put $y = \log T \log_2 T/(4B\log_3 T)$ for some $B \ge 5$, and $\delta = 1/[\log_2 T]^4$. Let $\| z \|$ denote the distance of z from the nearest integer.

Lemma 6. *For any real t_0 there is a positive integer $m \le T^{\frac{1}{B}}$ such that for each prime $p \le y$ we have $\| (mt_0\log p)/2\pi \| \le \delta$.*

Proof: This follows from Dirichlet's theorem on Diophantine approximation (see for example §8.2 of [8]) since $1/\delta$ is an integer and $(1/\delta)^{\pi(y)} \le T^{\frac{1}{B}}$, by the prime number theorem.

□

Lemma 7. *For any real t_1 there is a positive integer $n \le [\log_2 T]^2$ for which*

$$\mathrm{Re} \sum_{y \le p \le \exp((\log T)^{10})} \frac{1}{p^{1+int_1}} \ge -\frac{10}{\log_2 T}.$$

Proof: Let $K(x) = \max(0, 1 - |x|)$ and note that $\sum_{l=-L}^{L} K(l/L)e^{ilt}$ (the Fejer kernel) is non-negative for all positive integers L and all t. It follows therefore that

$$\sum_{j=-[\log_2 T]^2}^{[\log_2 T]^2} K\left(\frac{j}{[\log_2 T]^2}\right) \sum_{y \le p \le \exp((\log T)^{10})} \frac{1}{p^{1+ijt_1}} \ge 0.$$

Hence we obtain that

$$\begin{aligned}\mathrm{Re} \sum_{j=1}^{[\log_2 T]^2} K\left(\frac{j}{[\log_2 T]^2}\right) \sum_{y \le p \le \exp((\log T)^{10})} \frac{1}{p^{1+ijt_1}} &\ge -\frac{1}{2} \sum_{y \le p \le \exp((\log T)^{10})} \frac{1}{p} \\ &\ge -5 \log_2 T.\end{aligned}$$

The lemma follows at once. □

Proof of Theorem 2 : For $T^{\frac{1}{10}} \le |t| \le T$ one has

$$\log \zeta(1 + it) = - \sum_{p \le \exp((\log T)^{10})} \log\left(1 - \frac{1}{p^{1+it}}\right) + O\left(\frac{1}{\log T}\right).$$

(One can prove this, arguing as in the proof of the prime number theorem, by noting that $(1/2i\pi)\int_{(c)} \log \zeta(1 + it + w)(x^w/w)dw$ with $x = \exp((\log T)^{10})$ and $c > 0$ gives the main term of the right side by Perron's formula, and by shifting the contour to the left of 0, but enclosing a region free of zeros of $\zeta(s)$, we get residue $\log \zeta(1 + it)$ from the simple pole at $w = 0$, and the error term from the remaining integral.)

Combining Lemmas 6 and 7 (with $t_1 = mt_0$) we see that for any $t_0 \in [T^{1/10}, T]$ there exists an integer ℓ (where $\ell = mn$) with $1 \le \ell \le T^{\frac{1}{B}}[\log_2 T]^2$ such that $\| (\ell t_0 \log p)/2\pi \| \le 1/[\log_2 T]^2$ for each prime $p \le y$, and such that

$$\mathrm{Re} \sum_{y \le p \le \exp((\log T)^{10})} \frac{1}{p^{1+i\ell t_0}} \ge -\frac{10}{\log_2 T}.$$

We deduce therefore that

$$\begin{aligned}|\zeta(1 + i\ell t_0)| &\ge \prod_{p \le y} \left(1 - \frac{1}{p} + O\left(\frac{1}{p(\log_2 T)^2}\right)\right)^{-1} \left(1 + O\left(\frac{1}{\log_2 T}\right)\right) \\ &\ge e^{\gamma}(\log_2 T + \log_3 T - \log_4 T - \log A + O(1)),\end{aligned}$$

using the prime number theorem, where $A = 1/(2/B + 3 \log_2 T / \log T)$.

We use the above procedure with $t_0 = T_0, T_0 + 1, T_0 + 2, \ldots, T_0 + U_0$ where $T_0 = [T^{1-1/B}/3[\log_2 T]^2]$ and $U_0 = [T^{1-2/B}/7[\log_2 T]^4]$. Let ℓ_i be as above so $\ell_i \le T^{1/B}[\log_2 T]^2$

and thus $\tau_i = \ell_i(T_0 + i) \le T/2$. We claim that $|\tau_i - \tau_j| \ge 1$ if $i \ne j$ for if not then evidently $\ell_i \ne \ell_j$ (else $1 \le |(T_0 + j) - (T_0 + i)| = |\tau_j - \tau_i|/\ell_i < 1$), so that

$$T_0 \le |(\ell_i - \ell_j)T_0| \le |\tau_i - \tau_j| + |i\ell_i - j\ell_j| < 1 + U_0 T^{1/B}[\log_2 T]^2,$$

which is false. Now each $|\zeta(1 + i\tau_j)| \ge e^\gamma(\log_2 T + \log_3 T - \log_4 T - \log A + O(1))$. Since $|\zeta'(1 + it)| \ll \log^2 T$ for $1 \le |t| \le T$ we see that for any $|\alpha| \le 1/\log^2 T$ we have that $|\zeta(1 + i\tau_j + i\alpha)| = |\zeta(1 + i\tau_j)| + O(\alpha \log^2 T) = |\zeta(1 + i\tau_j)| + O(1)$. Thus the measure of $t \in [0, T]$ with $|\zeta(1 + it)| \ge e^\gamma(\log_2 T + \log_3 T - \log_4 T - \log A + O(1))$ is at least $2U_0/\log^2 T$, proving Theorem 2. □

7 The analogous results for L-functions at 1

By analogous methods one can prove:

Theorem 4. *Let q be a large prime.*

(i) *The proportion of characters $\chi \pmod q$ for which $|L(1,\chi)| > e^\gamma \tau$ is*

$$\exp\left(-\frac{2e^{\tau - C - 1}}{\tau}\left(1 + O\left(\frac{1}{\tau^{\frac{1}{2}}} + \left(\frac{e^\tau}{\log q}\right)^{\frac{1}{2}}\right)\right)\right), \tag{14}$$

uniformly in the range $1 \ll \tau \le \log_2 q - 20$. The same asymptotic also holds for the proportion of characters $\chi \pmod q$ for which $|L(1,\chi)| < \pi^2/6e^\gamma \tau$.

(ii) *There are at least $q^{1-1/A}$ characters $\chi \pmod q$ such that*

$$|L(1,\chi)| \ge e^\gamma(\log_2 q + \log_3 q - \log_4 q - \log A + O(1)),$$

for any $A \ge 10$.

If, in addition, we vary over all characters $\chi \pmod q$ and all primes $Q \le q \le 2Q$, then we can get a good estimate for the distribution function of $|L(1,\chi)|$ in almost the entire viable range. Thus we may prove that the proportion of $|L(1,\chi)| \ge e^\gamma\tau$ is (14) for the range $1 \le \tau \le \log_2 Q + \log_3 Q - 100$, but now with the error term "$(e^\tau/(\log Q \log_2 Q))^{\frac{1}{2}}$" in place of "$(e^\tau/\log q)^{\frac{1}{2}}$" (and a corresponding result holds for $1/|(6/\pi^2)L(1,\chi)|$).

The broad outline of the proof is the same, though now replacing $\log T$ by $\log Q \log_2 Q$, so that $\log Q(\log_2 Q)^4 \ge y \ge e^2 \log Q \log_2 Q$ and the range for k becomes $2 \le k \le \log Q \log_2 Q/(e^{10}\log(y/(\log Q \log_2 Q)))$. The result follows easily from the following analogy to Theorem 3,

$$\frac{1}{\pi(Q)}\sum_{q \le Q}\frac{1}{\varphi(q)}\sum_{\chi \pmod q}|L(1,\chi;y)|^{2z} = \prod_{p \le k}\left(1 - \frac{\delta}{p}\right)^{-2k\delta} \times \exp\left(\frac{2k}{\log k}\left(C_1 + O\left(\frac{k}{y} + \frac{1}{\log k}\right)\right)\right),$$

and an appropriate development of Lemma 4, where $L(1,\chi;y) := \prod_{p \le y}(1 - \chi(p)/p)^{-1}$. The above estimate, though, is proved rather more easily than Theorem 3. Since $L(1,\chi;y)^z = \sum_{n \in S(y)} d_z(n)$

$\chi(n)/n$, and $L(1,\overline{\chi};y)^z = \sum_{m\in S(y)} d_z(m)\overline{\chi}(m)/m$ where $S(y)$ is the set of integers all of whose prime factors are $\leq y$, the left side of this equation equals

$$\sum_{m,n\in S(y)} \frac{d_z(m)d_z(n)}{mn}\left\{\frac{1}{\pi(Q)}\sum_{q\leq Q}\frac{1}{\varphi(q)}\sum_{\chi \pmod q}\chi(m)\overline{\chi}(n)\right\}.$$

The term in {} equals $1-\#\{q\leq Q:\ q|mn\}/\pi(Q)$ if $m=n$, and is $\leq \#\{q\leq Q:\ q|m-n\}/\pi(Q)$ if $m\neq n$. Therefore our sum is

$$\sum_{n\in S(y)}\frac{d_z(n)^2}{n^2}+O\left(\frac{1}{\pi(Q)}\left(\sum_{m\in S(y)}\frac{|d_z(m)|\log 2m}{m}\right)^2\right).$$

Now $\log 2n \ll k^2+n^{1/k}$ so that

$$\begin{aligned}\sum_{n\in S(y)}\frac{|d_z(n)|}{n}\log 2n &\ll k^2\prod_{p\leq y}\left(1-\frac{\delta}{p}\right)^{-\delta k}+\prod_{p\leq y}\left(1-\frac{\delta}{p^{1-1/k}}\right)^{-\delta k}\\ &\ll \prod_{p\leq y}\left(1-\frac{\delta}{p}\right)^{-\delta k}\left(k^2+\exp\left(O\left(k\sum_{p\leq y}\frac{p^{1/k}-1}{p}\right)\right)\right)\\ &\ll (\log Q)^{O(1)}\prod_{p\leq y}\left(1-\frac{\delta}{p}\right)^{-\delta k},\end{aligned}$$

and the claimed estimate follows from Lemma 4.

References

[1] R. Balasubramanian, K. Ramachandra, and A. Sankaranarayanan, On the frequency of Titchmarsh's phenomenon for $\zeta(s)$-VIII, *Proc. Ind. Acad. Sci.* **102** (1992), 1-12.

[2] A. Granville and K. Soundararajan, The distribution of values of $L(1,\chi_d)$, *Geometric and Funct. Anal.* **13** (2003), 992-1028.

[3] N. Levinson, Ω-theorems for the Riemann zeta-function, *Acta Arith.* **20** (1972) 319–332.

[4] J.E. Littlewood, On the function $1/\zeta(1+it)$, *Proc. London Math. Soc.*, **27** (1928) 349-357.

[5] J.E. Littlewood, On the class number of the corpus $P(\sqrt{-k})$, *Proc. London Math. Soc.*, **27** (1928) 358-372.

[6] K. Ramachandra, On the frequency of Titchmarsh's phenomenon for $\zeta(s)$- VII, *Ann. Acad. Sci. Fenn.* **14** (1989) 27-40.

[7] G. Tenenbaum. *Introduction to analytic and probabilistic number theory*, Cambridge Studies in Advanced Mathematics **46**, Cambridge University Press, Cambridge (1995).

[8] E.C. Titchmarsh, *The theory of the Riemann zeta-function*, Oxford University Press, Oxford; Second edition, revised by D.R. Heath-Brown (1986).

[9] I.M. Vinogradov, A new estimate for $\zeta(1+it)$, *Izv. Akad. Nauk SSSR Ser. Mat.* **22** (1958) 161–164.

A. Granville
Départment de Mathématiques
et Statistique,
Université de Montréal,
CP 6128 succ Centre-Ville,
Montréal, QC H3C 3J7, Canada.

email: andrew@dms.umontreal.ca

K. Soundararajan
Department of Mathematics,
University of Michigan,
Ann Arbor, Michigan 48109, USA.

email: ksound@umich.edu

The Riemann Zeta Function and Related Themes – 2006, pp. 81–97

On moments of $|\zeta(\frac{1}{2} + it)|$ in short intervals

Aleksandar Ivić

Dedicated to Professor K. Ramachandra on his 70th birthday

Abstract

Power moments of

$$J_k(t, G) = \frac{1}{\sqrt{\pi}G} \int_{-\infty}^{\infty} |\zeta(\tfrac{1}{2} + it + iu)|^{2k} e^{-(u/G)^2} \, du \quad (t \asymp T, T^{\varepsilon} \le G \ll T),$$

where k is a natural number, are investigated. The results that are obtained are used to show how bounds for $\int_0^T |\zeta(\frac{1}{2} + it)|^{2k} \, dt$ may be obtained.

1 Introduction

Power moments represent one of the most important parts of the theory of the Riemann zeta-function $\zeta(s) = \sum_{n=1}^{\infty} n^{-s}$ $(\sigma = \Re e \, s > 1)$. Of particular significance are the moments on the "critical line" $\sigma = \frac{1}{2}$, and a vast literature exists on this subject (see e.g., [8], [9], [20], [22], [24] and [26]). Let us define

$$I_k(T) = \int_0^T |\zeta(\tfrac{1}{2} + it)|^{2k} \, dt, \tag{1}$$

where $k \in \mathbb{R}$ is a fixed, positive number. Naturally one would want to find an asymptotic formula for $I_k(T)$ for a given k, but this is an extremely difficult problem. Except when $k = 1$ and $k = 2$, no asymptotic formula for $I_k(T)$ is known yet, although there are plausible conjectures for such formulas (see e.g., [2]). In the absence of asymptotic formulas for $I_k(T)$, one would like then to obtain upper and lower bounds for $I_k(T)$, and for the closely related problem of

$$I_k(T + G) - I_k(T - G) = \int_{T-G}^{T+G} |\zeta(\tfrac{1}{2} + it)|^{2k} \, dt \quad (1 \ll G \le T). \tag{2}$$

For the latter, important results were obtained by K. Ramachandra, either alone, or in collaboration with R. Balasubramanian. Many of his results are contained in his comprehensive monograph [24] on mean values and omega-results for the Riemann zeta-function. In particular, [24] contains the proof of the lower bound

$$\int_{T-G}^{T+G} |\zeta(\tfrac{1}{2} + it)|^{2k} \, dt \gg_k G(\log G)^{k^2} \quad (\log\log T \ll_k G \le T, k \in \mathbb{N}), \tag{3}$$

2000 *Mathematics Subject Classification.* 11M06.

Key words and phrases. Riemann zeta-function, mean square and fourth moment of $|\zeta(\frac{1}{2} + it)|$, moments of $|\zeta(\frac{1}{2} + it)|$ in short intervals.

where $\ll_k$ (or $\gg_k$) means that the implied constant depends only on k. One believes that the bound in (3) represents the correct order of magnitude, at least for a certain range of G for a given $k \in \mathbb{N}$. Unfortunately, even proving the corresponding much weaker upper bound (for $G = T$), namely

$$I_k(T) \ll_{\varepsilon,k} T^{1+\varepsilon} \qquad (k > 0) \tag{4}$$

seems at present impossible for any $k > 2$. Here and later, $\varepsilon > 0$ denotes constants which may be arbitrarily small, but are not necessarily the same ones at each occurrence. In view of the relation (see [9] or [24])

$$\zeta^k(\tfrac{1}{2} + it) \ll_k \log t \left(\int_{t-1/3}^{t+1/3} |\zeta(\tfrac{1}{2} + iu)|^k \, dt \right) + 1 \qquad (k \in \mathbb{N}), \tag{5}$$

it is easily seen that (4) (for all k) is equivalent to the famous *Lindelöf hypothesis* that $\zeta(\frac{1}{2} + it) \ll_\varepsilon |t|^\varepsilon$. The Lindelöf hypothesis, like the even more famous *Riemann hypothesis* (that all complex zeros of $\zeta(s)$ have real part 1/2), is neither proved nor disproved at the time of writing of this text. For a discussion on this subject, see [13].

The aim of this paper is to investigate upper bounds for $I_k(T)$ when $k \in \mathbb{N}$, which we henceforth assume. The problem can be reduced to bounds of $|\zeta(\frac{1}{2} + it)|$ over short intervals, as in (2), but it is more expedient to work with the smoothed integral

$$J_k(T, G) := \frac{1}{\sqrt{\pi} G} \int_{-\infty}^{\infty} |\zeta(\tfrac{1}{2} + iT + iu)|^{2k} e^{-(u/G)^2} \, du \quad (1 \ll G \ll T). \tag{6}$$

Namely we obviously have

$$I_k(T + G) - I_k(T - G) = \int_{-G}^{G} |\zeta(\tfrac{1}{2} + iT + iu)|^{2k} \, du \le \sqrt{\pi} e G \, J_k(T, G), \tag{7}$$

and it is technically more convenient to work with $J_k(T, G)$ than with $I_k(T + G) - I_k(T - G)$. Of course, instead of the Gaussian exponential weight $\exp(-(u/G)^2)$, one could introduce in (6) other smooth weights with a similar effect. The Gaussian weight has the advantage that, by the use of the classical integral

$$\int_{-\infty}^{\infty} \exp(Ax - Bx^2) \, dx = \sqrt{\frac{\pi}{B}} \exp\left(\frac{A^2}{4B}\right) \qquad (\Re e\, B > 0), \tag{8}$$

one can often explicitly evaluate the relevant exponential integrals that appear in the course of the proof.

The plan of the paper is as follows. In the next two sections we shall briefly discuss the results on $I_k(T)$ and $J_k(T, G)$ when $k = 1$ and $k = 2$, respectively. Indeed, as these are the only cases when we possess relatively good knowledge and explicit formulas, it is only natural that those results be used in deriving results on higher power moments, when our knowledge is quite imperfect. We shall obtain new results on moments of $J_k(T, G)$ by using the explicit formulas of Section 2 and Section 3. This will be done in Section 4 and Section 5. Finally, in Section 6, it will be shown how one can obtain bounds for $I_k(T)$ from the bounds of moments of $J_k(T, G)$.

2 The mean square formula

The mean square formula for $|\zeta(\frac{1}{2}+it)|$ is traditionally written in the form

$$\int_0^T |\zeta(\tfrac{1}{2}+it)|^2\,dt = T\log\left(\frac{T}{2\pi}\right) + (2\gamma-1)T + E(T), \tag{9}$$

where $\gamma = -\Gamma'(1) = 0.577\ldots$ is Euler's constant, and $E(T)$ is to be considered as the error term in the asymptotic formula (9). F.V. Atkinson [1] established in 1949 an explicit, albeit complicated formula for $E(T)$, containing two exponential sums of length $\asymp T$ weighted by the number of divisors function $d(n)$, plus an error term which is $O(\log^2 T)$. This is given as

Lemma 1. *Let* $0 < A < A'$ *be any two fixed constants such that* $AT < N < A'T$, *and let* $N' = N'(T) = T/(2\pi) + N/2 - (N^2/4 + NT/(2\pi))^{1/2}$. *Then*

$$E(T) = \Sigma_1(T) + \Sigma_2(T) + O(\log^2 T), \tag{10}$$

where

$$\Sigma_1(T) = 2^{1/2}(T/(2\pi))^{1/4}\sum_{n\le N}(-1)^n d(n) n^{-3/4} e(T,n)\cos(f(T,n)), \tag{11}$$

$$\Sigma_2(T) = -2\sum_{n\le N'} d(n) n^{-1/2}(\log T/(2\pi n))^{-1}\cos(T\log T/(2\pi n) - T + \pi/4), \tag{12}$$

with

$$\begin{aligned} f(T,n) &= 2T\,\mathrm{arsinh}\left(\sqrt{\pi n/(2T)}\right) + \sqrt{2\pi nT + \pi^2 n^2} - \tfrac{1}{4}\pi \\ &= -\tfrac{1}{4}\pi + 2\sqrt{2\pi nT} + \tfrac{1}{6}\sqrt{2\pi^3}\, n^{3/2} T^{-1/2} \\ &\quad + a_5 n^{5/2} T^{-3/2} + a_7 n^{7/2} T^{-5/2} + \cdots, \end{aligned} \tag{13}$$

$$\begin{aligned} e(T,n) &= (1+\pi n/(2T))^{-1/4}\left\{(2T/\pi n)^{1/2}\mathrm{arsinh}\left(\sqrt{\pi n/(2T)}\right)\right\}^{-1} \\ &= 1 + O(n/T) \qquad (1 \le n < T), \end{aligned} \tag{14}$$

and $\mathrm{arsinh}\,x = \log(x + \sqrt{1+x^2})$.

Atkinson's formula was the starting point for many results on $E(T)$ (see [8, Chapter 15] for some of them). It is conjectured that $E(T) \ll_\varepsilon T^{1/4+\varepsilon}$, but currently this bound cannot be proved even if the Riemann Hypotheis is assumed. The best known upper bound for $E(T)$, obtained by intricate estimation of a certain exponential sum, is due to M.N. Huxley [6]. This is

$$E(T) \ll T^{72/227}(\log T)^{679/227}, \qquad \frac{72}{227} = 0.3171806\ldots.$$

In the other direction, J.L. Hafner and the author [3] proved that there exist absolute constants $A, B > 0$ such that

$$E(T) = \Omega_+\left\{(T\log T)^{1/4}(\log\log T)^{(3+\log 4)/4} e^{-A\sqrt{\log\log\log T}}\right\}$$

and

$$E(T) = \Omega_-\left\{T^{1/4}\exp\left(\frac{B(\log\log T)^{1/4}}{(\log\log\log T)^{3/4}}\right)\right\},$$

where $f(x) = \Omega_+(g(x))$ means that $\limsup_{x\to\infty} f(x)/g(x) > 0$, and $f(x) = \Omega_-(g(x))$ means that $\liminf_{x\to\infty} f(x)/g(x) < 0$.

Higher moments of $E(t)$ were investigated by K.-M. Tsang [27], who proved that, for some $\beta_1 < 7/4$,

$$\int_1^T E^3(t)\,\mathrm{d}t = B_1 T^{7/4} + O_\varepsilon(T^{\beta_1+\varepsilon}) \qquad (B_1 > 0) \tag{15}$$

and for some $\gamma_1 < 2$

$$\int_1^T E^4(t)\,\mathrm{d}t = C_1 T^2 + O_\varepsilon(T^{\gamma_1+\varepsilon}) \qquad (C_1 > 0), \tag{16}$$

which supports the conjecture that $E(T) \ll_\varepsilon T^{1/4+\varepsilon}$. The values $\beta_1 = 5/3$ and $\gamma_1 = 23/12$ are obtained in the recent work of P. Sargos and the author [18].

In what follows we shall formulate an explicit formula for $J_1(T,G)$. Such a result can be, of course, deduced from Atkinson's formula (11)-(13) by the use of (8). This approach was used originally by D.R. Heath-Brown [4], who proved

$$\int_0^T |\zeta(\tfrac{1}{2}+it)|^{12}\,\mathrm{d}t \ll T^2 \log^{17} T, \tag{17}$$

which is still essentially the best known result concerning higher power moments of $|\zeta(\frac{1}{2}+it)|$. This procedure can be avoided by appealing to Y. Motohashi's formula [22, p. 213], which states that

$$J_1(T,G) = 2^{\frac{3}{4}}\pi^{\frac{1}{4}}T^{-\frac{1}{4}}\sum_{n=1}^{\infty}(-1)^n d(n) n^{-\frac{3}{4}} \sin f(T,n)$$
$$\exp\Big(-\frac{\pi n G^2}{2T}\Big) + O(\log T), \tag{18}$$

where $f(T,n)$ is given by (13), and $T^{1/4} \le G \le T/\log T$. In fact, only the range $G \le T^{1/3}$ is relevant, since for $G \ge T^{1/3}$ one has $J_1(T,G) \ll \log T$ by [8, Chapter 7]. Motohashi's proof of (18), like the proof of Atkinson's formula for $E(T)$, is based on classical methods from analytic number theory. Albeit the expression on the right-hand side of (18) is quite simple, the condition $G \ge T^{1/4}$ is rather restrictive for the application that we have in mind. Thus we shall use a similar type of result, which is valid in a much wider range. This is contained in

Lemma 2. *For $T^\varepsilon \le G \le T$ and $f(T,n)$ given by* (13), *we have*

$$J_1(T,G) = O(\log T) +$$
$$+\sqrt{2}\sum_{n=1}^{\infty}(-1)^n d(n) n^{-1/2}\bigg(\Big(\frac{T}{2\pi n}+\frac{1}{4}\Big)^{1/2}-\frac{1}{2}\bigg)^{-1/2} \times \tag{19}$$
$$\times \exp\Big(-G^2\big(\operatorname{arsinh}\sqrt{\pi n/(2T)}\big)^2\Big)\sin f(T,n).$$

By using Taylor's formula it is seen that the error made by replacing

$$\bigg(\Big(\frac{T}{2\pi n}+\frac{1}{4}\Big)^{1/2}-\frac{1}{2}\bigg)^{-1/2} \exp\Big(-G^2\big(\operatorname{arsinh}\sqrt{\pi n/(2T)}\big)^2\Big)$$

by

$$\left(\frac{T}{2\pi n}\right)^{-1/4}\exp(-\pi nG^2/(2T))$$

is $\ll 1$ for $G \ge T^{1/5}\log^C T$. But the important fact is that in applications (19) is as useful as (18), since the factors under the sine function are identical.

Proof of Lemma 2: The proof of (19) follows from Y. Motohashi [22, Theorem 4.1], which gives that (18)

$$\int_{-\infty}^{\infty}|\zeta(\tfrac{1}{2}+it)|^2 g(t)\,\mathrm{d}t = \int_{-\infty}^{\infty}\left[\Re\mathrm{e}\left\{\frac{\Gamma'}{\Gamma}(\tfrac{1}{2}+it)\right\}+2\gamma-\log(2\pi)\right]g(t)\,\mathrm{d}t$$
$$+\,2\pi\Re\mathrm{e}\,(g(\tfrac{1}{2}i)) + 4\sum_{n=1}^{\infty}d(n)\int_0^{\infty}(y(y+1))^{-1/2}g_c(\log(1+1/y))\cos(2\pi ny)\,\mathrm{d}y,$$

where

$$g_c(x) := \int_{-\infty}^{\infty} g(t)\cos(xt)\,\mathrm{d}t$$

is the cosine Fourier transform of $g(t)$. One requires the function $g(r)$ to be real-valued for $r \in \mathbb{R}$, and that there exists a large constant $A > 0$ such that $g(r)$ is regular and $\ll (|r|+1)^{-A}$ for $|\Im\mathrm{m}\, r| \le A$. The choice

$$g(t) = \frac{1}{\sqrt{\pi}G}\mathrm{e}^{-(T-t)^2/G^2}, \quad g_c(x) = \mathrm{e}^{-\frac{1}{4}(Gx)^2}\cos(Tx)$$

is permissible, and then the integral on the left-hand side of (18) becomes $J_1(T,G)$. The first integral on the right-hand side of (18) is $O(\log T)$, and the second one is evaluated by the saddle-point method (see e.g., [8, Chapter 2]). A convenient result to use is [8, Theorem 2.2 and Lemma 15.1], due originally to Atkinson [1]. In the latter only the exponential factor $\exp(-\frac{1}{4}G^2\log(1+1/y))$ is missing. In the notation of [1] and [8] we have that the saddle point x_0 satisfies

$$x_0 = U - \frac{1}{2} = \left(\frac{T}{2\pi n}+\frac{1}{4}\right)^{1/2} - \frac{1}{2},$$

and the presence of the above exponential factor makes it possible to truncate the series in (19) at $n = TG^{-2}\log T$ with a negligible error. Furthermore, in the remaining range for n we have

$$\Phi_0\mu_0F_0^{-3/2} \ll (nT)^{-3/4},$$

which makes a total copntribution of $O(1)$, as does error term integral in Theorem 2.2 of [8]. The error terms with $\Phi(a)$, $\Phi(b)$ vanish for $a = 0$, $b = \infty$, and (19) follows □

3 The formula for the fourth moment

The asymptotic formula for the fourth moment of the Riemann zeta-function $\zeta(s)$ on the critical line is customarily written as

$$\int_0^T |\zeta(\tfrac{1}{2}+it)|^4\,\mathrm{d}t = TP_4(\log T) + E_2(T),\ P_4(x) = \sum_{j=0}^{4}a_jx^j. \tag{20}$$

A classical result of A.E. Ingham [7] from 1926 is that $a_4 = 1/(2\pi^2)$ and that the error term $E_2(T)$ in (20) satisfies the bound $E_2(T) \ll T\log^3 T$ (a simple proof of this is due to K. Ramachandra [23]). Much later D.R. Heath-Brown [4] made progress in this problem by proving that $E_2(T) \ll_\varepsilon T^{7/8+\varepsilon}$. He also calculated

$$a_3 = 2(4\gamma - 1 - \log(2\pi) - 12\zeta'(2)\pi^{-2})\pi^{-2}$$

and produced more complicated expressions for a_0, a_1 and a_2 in (20). For an explicit evaluation of the a_j's in (20) the reader is referred to the author's work [11]. In the last fifteen years, due primarily to the application of powerful methods of spectral theory (see Y. Motohashi's monograph [22] for a comprehensive account), much advance has been made in connection with $E_2(T)$. This involves primarily results with exponential sums involving the quantities κ_j and $\alpha_j H_j^3(\frac{1}{2})$. Here as usual $\{\lambda_j = \kappa_j^2 + \frac{1}{4}\} \cup \{0\}$ is the discrete spectrum of the non-Euclidean Laplacian acting on $SL(2,\mathbb{Z})$–automorphic forms, and $\alpha_j = |\rho_j(1)|^2(\cosh \pi\kappa_j)^{-1}$, where $\rho_j(1)$ is the first Fourier coefficient of the Maass wave form corresponding to the eigenvalue λ_j to which the Hecke series $H_j(s)$ is attached. It is conjectured that $E_2(T) \ll_\varepsilon T^{1/2+\varepsilon}$, which would imply the (hitherto unproved) bound $\zeta(\frac{1}{2} + it) \ll_\varepsilon |t|^{1/8+\varepsilon}$. It is known now that

$$E_2(T) = O(T^{2/3}\log^{C_1} T), \quad E_2(T) = \Omega(T^{1/2}), \tag{21}$$

$$\int_0^T E_2(t)\,dt = O(T^{3/2}), \quad \int_0^T E_2^2(t)\,dt = O(T^2\log^{C_2} T), \tag{22}$$

with effective constants $C_1, C_2 > 0$ (the values $C_1 = 8$, $C_2 = 22$ are worked out in [22]). The above results were proved by Y. Motohashi and the author: (21) and the first bound in (22) in [17], and the second upper bound in (22) in [16]. The Ω–result in (21) was improved to $E_2(T) = \Omega_\pm(T^{1/2})$ by Y. Motohashi [21]. It turns out that there is no explicit formula for $E_2(T)$ which would represent the analogue of Atkinson's formula (cf. Lemma 1). Results on $E_2(T)$ have been obtained indirectly, by using the explicit formula for $J_2(T,G)$, due to Y. Motohashi (see [22]). This is

Lemma 3. *Let $D > 0$ be an arbitrary constant. For $T^{1/2}\log^{-D} T \le G \le T/\log T$ we have*

$$J_2(T,G) = O(\log^{3D+9} T) + \frac{\pi}{\sqrt{2T}}\sum_{j=1}^\infty \alpha_j H_j^3(\tfrac{1}{2})\kappa_j^{-1/2}\sin\left(\kappa_j\log\frac{\kappa_j}{4eT}\right)\exp(-\tfrac{1}{4}(G\kappa_j/T)^2). \tag{23}$$

In what concerns higher moments, let us only state that from (17) and (20) one obtains by Hölders's inequality for integrals

$$\int_0^T |\zeta(\tfrac{1}{2} + it)|^6\,dt \ll T^{5/4}\log^{29/4} T, \quad \int_0^T |\zeta(\tfrac{1}{2} + it)|^8\,dt \ll T^{3/2}\log^{21/2} T. \tag{24}$$

The bounds in (24) are hitherto the sharpest ones known.

Let it be also mentioned here that the sixth moment was investigated by the author in [12], where it was shown that

$$\int_0^T |\zeta(\tfrac{1}{2} + it)|^6\,dt \ll_\varepsilon T^{1+\varepsilon} \tag{25}$$

does hold if a certain conjecture involving the so-called ternary additive divisor problem is true.

4 The moments of $J_1(t,G)$

In this section we shall prove results on moments of $J_1(t,G)$. One expects this function, at least for certain ranges of G, to behave like $O(t^\varepsilon)$ on the average. Our bounds are contained in

Theorem 1. *We have*

$$\int_T^{2T} J_1^m(t,G)\,\mathrm{d}t \ll_\varepsilon T^{1+\varepsilon} \tag{26}$$

for $T^\varepsilon \le G \le T$ *if* $m = 1,2$; *for* $T^{1/7+\varepsilon} \le G \le T$ *if* $m = 3$, *and for* $T^{1/5+\varepsilon} \le G \le T$ *if* $m = 4$.

Proof of Theorem 1: Our starting point in all cases is the explicit formula (19). The results for $m = 1$ and $m = 2$ follow by straightforward integration and the first derivative test for exponential integrals (see [8, Lemma 2.1]). The proof resembles mean square bounds for $\Delta(x)$ (the error term in the divisor problem) and $E(t)$ (op. cit.), and is omitted for the sake of brevity. Instead, we shall concentrate on the more difficult cases $m = 3$ and $m = 4$. For this we shall need two lemmas on the spacing of three and four square roots (the square roots appear in view of the asymptotic formula given in (13)). These are □

Lemma 4. *Let N denote the number of solutions in integers m, n, k of the inequality*

$$\left|\sqrt{m} + \sqrt{n} - \sqrt{k}\right| \le \delta\sqrt{M} \qquad (\delta > 0)$$

with $M' < n \le 2M'$, $M < m \le 2M$, and $M' \le M$. Then

$$N \ll_\varepsilon M^\varepsilon(M^2M'\delta + (MM')^{1/2}). \tag{27}$$

Lemma 5. *Let $k \ge 2$ be a fixed integer and $\delta > 0$ be given. Then the number of integers n_1, n_2, n_3, n_4 such that $N < n_1, n_2, n_3, n_4 \le 2N$ and*

$$\left|n_1^{1/k} + n_2^{1/k} - n_3^{1/k} - n_4^{1/k}\right| < \delta N^{1/k}$$

is, for any given $\varepsilon > 0$,

$$\ll_\varepsilon N^\varepsilon(N^4\delta + N^2). \tag{28}$$

Lemma 4 was proved by Sargos and the author [18], while Lemma 5 is due to Robert–Sargos [25]. The plan of the proof of (26) when $m = 3$ is simple: the expression (19) will be raised to the third power and then integrated. There are, however, two obstacles in attaining this goal. The first is that direct integration does not lead to adequate truncation, so that some smoothing of the relevant integral will be made. The second one is that in the asymptotic formula for $f(T,n)$ in (13) not only square roots appear, but also higher powers. To get around this difficulty we shall appeal to

Lemma 6. (M. Jutila [19]). *For $A \in \mathbb{R}$ we have*

$$\cos\left(\sqrt{8\pi nT} + \tfrac{1}{6}\sqrt{2\pi^3}n^{3/2}T^{-1/2} - A\right) = \int_{-\infty}^{\infty} \alpha(u)\cos(\sqrt{8\pi n}(\sqrt{T} + u) - A)\,\mathrm{d}u, \tag{29}$$

where $\alpha(u) \ll T^{1/6}$ for $u \ne 0$,

$$\alpha(u) \ll T^{1/6}\exp(-bT^{1/4}|u|^{3/2}) \tag{30}$$

for $u < 0$, and

$$\alpha(u) = T^{1/8}u^{-1/4}\left(d\exp(ibT^{1/4}u^{3/2}) + \bar{d}\exp(-ibT^{1/4}u^{3/2})\right) + O(T^{-1/8}u^{-7/4}) \tag{31}$$

for $u \ge T^{-1/6}$ and some constants b (> 0) and d.

Now we continue with the proof of Theorem 1. Write first

$$\int_T^{2T} J_1^m(t,G)\,dt \le \int_{T/2}^{5T/2} \varphi(t)J_1^m(t,G)\,dt, \tag{32}$$

where $\varphi(t)$ (≥ 0) is a smooth function supported in $[T/2, 5T/2]$ such that $\varphi(t) = 1$ when $t \in [T, 2T]$, and then we have $\varphi^{(r)}(t) \ll_r T^{-r}$ ($r = 0, 1, 2, \ldots$). We truncate (19) at $TG^{-2}\log T$ and use it to expand the m-th power on the right-hand side of (32) when $m = 3, 4$. The terms $a_{2j-1}n^{(2j-1)/2}T^{-j/2} (j \ge 3)$ in $f(T,n)$ (cf. (13)) are expanded by Taylor's formula. Since

$$n^{5/2}t^{-3/2} \ll TG^{-5}\log^{5/2} T \le T^{-\varepsilon}$$

for $G \ge T^{1/5+\varepsilon}$, it transpires that in this range for G we may take sufficiently many terms in Taylor's formula so that the error term will make a negligible contribution. The other terms will lead to similar expressions, and the largest contribution will come from the constant term. After this there will remain

$$\sin h(t,k)\sin h(t,\ell)\sin h(t,n),\ h(t,u) = \sqrt{8\pi tu} + \tfrac{1}{6}\sqrt{2\pi^3}u^{3/2}t^{-1/2} - \tfrac{1}{4}\pi$$

when $m = 3$, with $T^{1/3} < k, \ell, n \le TG^{-2}\log T$, $k \asymp K$, $k \ge \max(\ell, n)$. The factors with $u^{3/2}$ (for $u = k, \ell, n$) are removed from the h-functions by the use of Lemma 6. With $\alpha(v)$ given by (31) we have

$$\begin{aligned}&\cos\left(\sqrt{8\pi nt} + \tfrac{1}{6}\sqrt{2\pi^3}n^{3/2}t^{-1/2} - A\right) = O(T^{-10}) + \\ &\int_{-u_0}^{u_1} \alpha(v)\cos(\sqrt{8\pi n}(\sqrt{t}+v) - A)\,dv + \int_{u_1}^{\infty} \alpha(v)\cos(\sqrt{8\pi n}(\sqrt{t}+v) - A)\,dv,\end{aligned} \tag{33}$$

where we set

$$u_0 = T^{-1/6}\log T,\ u_1 = CKT^{-1/2}, \tag{34}$$

and $C > 0$ is a large constant. With this choice of u_0, u_1 and (30)-(31) it follows that, for $T/2 \le t \le 5T/2$,

$$\int_{-u_0}^{u_1} \alpha(v)\cos(\sqrt{8\pi n}(\sqrt{t}+v) - A)\,dv + \int_{u_1}^{\infty} \alpha(v)\cos(\sqrt{8\pi n}(\sqrt{t}+v) - A)\,dv \ll \log T. \tag{35}$$

Namely we have

$$\int_{u_0}^{u_1} t^{1/8}v^{-1/4}\exp(ibt^{1/4}v^{3/2} \pm \sqrt{8\pi nv})\,dv \ll \log T,$$

on writing the integral as a sum of $\ll \log T$ integrals over $[U, U']$ with $u_0 \le U < U' \le 2U \ll u_1$, and applying the second derivative test (i.e., [8, Lemma 2.2]) to each of these integrals.

We remark that the contribution of the O-term in (31) will be, by trivial estimation, $O(1)$. It remains yet to deal with the integral with $v > u_1$ in (35), when we note that

$$\frac{\partial}{\partial v}\left(bt^{1/4}v^{3/2} \pm \sqrt{8\pi n v}\right) \gg T^{1/4}v^{1/2} \qquad (v > u_1, t \asymp T),$$

provided that C in (34) is sufficiently large. Thus by the first derivative test

$$\begin{aligned}\int_{u_1}^{\infty} &\alpha(v)\cos(\sqrt{8\pi n}(\sqrt{t}+v) - A)\,\mathrm{d}v \\ &\ll 1 + T^{1/8}u_1^{-1/4}T^{-1/4}u_1^{-1/2} \\ &\ll 1 + T^{1/4}K^{-3/4} \ll 1,\end{aligned}$$

since $K \gg T^{1/3}$. Thus (35) holds.

Hence setting

$$E_{\pm} := \sqrt{8\pi}(\sqrt{k} + \sqrt{\ell} \pm \sqrt{n}),$$

it is seen that we are left with the integral of

$$\int_{T/2}^{5T/2} \varphi(t)F(t;k,\ell,n)\mathrm{e}^{iE_{\pm}\sqrt{t}}\,\mathrm{d}t, \tag{36}$$

where

$$\begin{aligned}F(t;k,\ell,n) := &\left(\left(\frac{t}{2\pi k}+\frac{1}{4}\right)^{1/2}-\frac{1}{2}\right)^{-1/2}\left(\left(\frac{t}{2\pi \ell}+\frac{1}{4}\right)^{1/2}-\frac{1}{2}\right)^{-1/2} \\ &\times\left(\left(\frac{t}{2\pi n}+\frac{1}{4}\right)^{1/2}-\frac{1}{2}\right)^{-1/2}\exp(-G^2(\operatorname{arsinh}\sqrt{\pi k/(2t)})^2) \\ &\times\exp(-G^2(\operatorname{arsinh}\sqrt{\pi \ell/(2t)})^2)\exp(-G^2(\operatorname{arsinh}\sqrt{\pi n/(2t)})^2).\end{aligned}$$

Repeated integration by parts show that the integral in (36) with E_+ will make a negligible contribution, and also the one with $|E_-| \ge T^{\varepsilon-1/2}$ for any given $\varepsilon > 0$. The contribution of those k, ℓ, n for which $|E_-| \le T^{\varepsilon-1/2}$ is estimated by the use of Lemma 4 (with an obvious change of notation and with $\delta \asymp K^{-1/2}T^{\varepsilon-1/2}$). After this, the integral over t is estimated trivially, and (35) is used. The relevant expression on the right-hand side of (32) is

$$\begin{aligned}&\ll_{\varepsilon} T^{1+\varepsilon} \max_{T^{1/3}\le K\le TG^{-2}\log T}(TK)^{-3/4}(K^3\cdot K^{-1/2}T^{-1/2}+K) \\ &\ll_{\varepsilon} T^{\varepsilon}\max_{K\le TG^{-2}\log T}(K^{7/4}T^{-1/4}+T^{1/4}K^{1/4}) \\ &\ll_{\varepsilon} T^{3/2+\varepsilon}G^{-7/2}+T^{1/2+\varepsilon} \ll_{\varepsilon} T^{1+\varepsilon}\end{aligned}$$

for $G \ge T^{1/7+\varepsilon}$. However, our initial condition was $G \ge T^{1/5+\varepsilon}$, which is more restrictive. Fortunately, this is a technical point that can be resolved by modifying Lemma 6 suitably. Namely, instead of $n^{3/2}T^{-1/2}$ in (29) we may put $Cn^{5/2}T^{-3/2}$, which will be "removed" in the fashion of Lemma 6. Instead of the function $\alpha(u)$, another oscillating function $\beta(u)$ will appear, for which the analogue of (33) will hold. Jutila obtained (30)–(31) by exploiting the fact that the inversion made in (29) can be connected to the *Airy integral*

$$\mathrm{Ai}(x) := \frac{1}{\pi}\int_0^{\infty}\cos(\tfrac{1}{3}t^3+tx)\,\mathrm{d}t \qquad (x \ge 0),$$

for which there exist representations in terms of the classical Bessel functions, thereby providing quickly asymptotic expansions necessary in Lemma 6. In the new case there will be no Airy integrals involved, but the necessary asymptotic expansion can be obtained by the use of the saddle point method.

This ends the discussion of the case $m = 3$. The case $m = 4$ will be analogous, the non-trivial contribution will come from integer quadruples (n_1, n_2, n_3, n_4) such that $K < n_1, n_2, n_3, n_4 \le 2K$ $(T^{1/3} \le K \le TG^{-2}\log T)$ and

$$|\sqrt{n_1} + \sqrt{n_2} - \sqrt{n_3} - \sqrt{n_4}| \le T^{\varepsilon-1/2}. \tag{37}$$

Instead of Lemma 4 we use Lemma 5 (with $k = 2$), to obtain a contribution which is

$$\begin{aligned}
&\ll_\varepsilon T^{1+\varepsilon} \max_{K\le TG^{-2}\log T} (TK)^{-1}(K^4 \cdot K^{-1/2}T^{-1/2} + K^2)\\
&\ll_\varepsilon T^{\varepsilon} \max_{K\le TG^{-2}\log T} (T^{-1/2}K^{5/2} + K)\\
&\ll_\varepsilon T^{2+\varepsilon}G^{-5} + T^{1+\varepsilon} \ll_\varepsilon T^{1+\varepsilon}
\end{aligned}$$

for $G \ge T^{1/5+\varepsilon}$, as asserted. In this case direct application of Lemma 6 suffices. Values of m satisfying $m > 4$ in (26) could be handled in a similar fashion, provided that one can find analogues of Lemma 4 and Lemma 5 which are strong enough.

5 The moments of $J_2(t, G)$

We shall prove now the analogue of Theorem 1 for $J_2(t, G)$. This is a more difficult problem, and the ranges for G for which the analogue of (26) will hold will be poorer. The result is

Theorem 2. *We have*

$$\int_T^{2T} J_2^m(t, G)\,dt \ll_\varepsilon T^{1+\varepsilon} \tag{38}$$

for $T^{1/2+\varepsilon} \le G \le T$ *if* $m = 1, 2$; *for* $T^{4/7+\varepsilon} \le G \le T$ *if* $m = 3$, *and for* $T^{3/5+\varepsilon} \le G \le T$ *if* $m = 4$.

Proof of Theorem 2. Our starting point is Lemma 3. We remark that by using the estimate (see Y. Motohashi [22, Section 3.4])

$$\sum_{\kappa_j\le K} \alpha_j H_j^3(\tfrac12) \ll_\varepsilon K^2 \log^3 K \tag{39}$$

we see that for $G \ge T^{2/3}\log^C T$ the right-hand side of (23) is $O(1)$, hence we may suppose that $T^{1/2+\varepsilon} \le G \le T^{2/3}\log^C T$. Observe also that we may truncate the series in (23) at $TG^{-1}\log T$ with a negligible error. Besides (39) we need one more ingredient from spectral theory, namely the author's bound [14]

$$\sum_{K-1\le\kappa_j\le K+1} \alpha_j H_j^3(\tfrac12) \ll_\varepsilon K^{1+\varepsilon}. \tag{40}$$

We give now the proof of Theorem when $m = 2$ (the case $m = 1$ easily follows from this and the Cauchy-Schwarz inequality). We use (32) with J_2 replacing J_1. Then

$$\int_T^{2T} J_2^2(t,G)\,\mathrm{d}t \ll_\varepsilon T^{1+\varepsilon} + T^{\varepsilon-1} \max_{K\le TG^{-1}\log T} \sum_{K<\kappa_j,\kappa_\ell\le 2K} \alpha_j\alpha_l H_j^3(\tfrac12)H_\ell^3(\tfrac12)\times$$

$$\times (\kappa_j\kappa_\ell)^{-1/2}\left(\frac{\kappa_j}{4\mathrm{e}}\right)^{i\kappa_j}\left(\frac{\kappa_\ell}{4\mathrm{e}}\right)^{-i\kappa_\ell}\int_T^{2T}\varphi(t)t^{i\kappa_j-i\kappa_\ell}\exp\left(-\tfrac14 G^2t^{-2}(\kappa_j^2+\kappa_\ell^2)\right)\mathrm{d}t.$$

Repeated integrations by parts show that the contribution of κ_j, κ_ℓ for which $|\kappa_j - \kappa_\ell| \ge T^\varepsilon$ is negligible. The contribution of $|\kappa_j - \kappa_\ell| < T^\varepsilon$ is estimated by (40) (splitting the summation over κ_ℓ in subsums of length ≤ 2) and (39), while the integral over t is estimated trivially. The contribution will be

$$\begin{aligned}
&\ll_\varepsilon T^{1+\varepsilon}\max_{K\le TG^{-1}\log T} T^{-1}\sum_{K<\kappa_j\le 2K}\alpha_j H_j^3(\tfrac12)\kappa_j^{-1/2}\sum_{|\kappa_j-\kappa_\ell|<T^\varepsilon}\alpha_l H_\ell^3(\tfrac12)\kappa_\ell^{-1/2}\\
&\ll_\varepsilon T^\varepsilon \max_{K\le TG^{-1}\log T}\sum_{K<\kappa_j\le 2K}\alpha_j H_j^3(\tfrac12)K^{-1/2}K^{1/2}\\
&\ll_\varepsilon T^{2+\varepsilon}G^{-2} \ll_\varepsilon T^{1+\varepsilon}
\end{aligned}$$

for $G \ge T^{1/2+\varepsilon}$, as asserted. We remark that the technique of this proof can be used, following the arguments in [9, Chapter 5], to yield a quick proof of the important bound

$$\int_0^T E_2^2(t)\,\mathrm{d}t \ll_\varepsilon T^{2+\varepsilon},$$

which is only slightly weaker than the second bound in (22).

The cases $m = 3$ and $m = 4$ are dealt with analogously. For the former, after we raise the sum in (23) to the cube, it is seen that the non-negligible contribution comes from the triplets $(\kappa_j, \kappa_m, \kappa_\ell)$ for which $|\kappa_j + \kappa_m - \kappa_\ell| < T^\varepsilon$. For the summation over one of the variables, say κ_ℓ, we use (40), and for the summation over κ_j, κ_m we use (39). Similarly, in the case of the fourth power the non-negligible contribution will come from quadruples $(\kappa_j, \kappa_m, \kappa_\ell, \kappa_n)$ for which

$$|\kappa_j + \kappa_m - \kappa_\ell - \kappa_n| < T^\varepsilon. \tag{41}$$

In this way the assertions of Theorem 2 concerning the cases $m = 3, 4$ are obtained; the details are omitted for the sake of brevity. Note that $\lambda_j = \kappa_j^2 + \frac14$, so that (41) can be rewritten as

$$|\sqrt{\lambda_j} + \sqrt{\lambda_m} - \sqrt{\lambda_\ell} - \sqrt{\lambda_n}| < T^\varepsilon, \tag{42}$$

which is somewhat analogous to (37). Lemma 5 provides a good bound for the number of integer quadruples satisfying (37), but the condition (42) is much more difficult to deal with, since little is known about arithmetic properties of the spectral values λ_j. □

6 Bounds for moments of $|\zeta(\frac12 + it)|$

We shall show now how the results on power moments of $|\zeta(\frac12 + it)|$ follow from mean square results on short intervals. In particular, a new result will be derived, which connects power moments of $|\zeta(\frac12 + it)|$ with upper bounds furnished by Theorem 1 and Theorem 2.

To begin with, suppose that $\{t_r\}_{r=1}^R$ are points lying in $[T, 2T]$ such that $t_{r+1} - t_r \geq 1$ $(r = 1, \ldots, R-1)$ and $|\zeta(\frac{1}{2} + it_r)| \geq V \geq T^\varepsilon$ for $r = 1, \ldots, R$. From (5) we have

$$\begin{aligned} RV^{2k} &\ll \log T \sum_{r=1}^{R} \int_{t_r-1/3}^{t_r+1/3} |\zeta(\tfrac{1}{2} + it)|^{2k}\, \mathrm{d}t \\ &\ll \log T \sum_{s=1}^{S} \int_{\tau_s-G}^{\tau_s+G} |\zeta(\tfrac{1}{2} + it)|^{2k}\, \mathrm{d}t \\ &\ll G \log T \sum_{s=1}^{S} J_k(\tau_s, G), \end{aligned} \tag{43}$$

where we have grouped integrals over disjoint intervals $[t_r - 1/3, t_r + 1/3]$ into integrals over disjoint intervals $[\tau_s - G, \tau_s + G]$ with $s = 1, \ldots, S (\leq R), G \geq T^\varepsilon$.

Suppose now that $k = 1$ in (43). Then we use Lemma 2, noting that there are no absolute value signs on the right-hand side of (19), which can be truncated at $TG^{-2} \log T$ and where, as before, we may assume that $G \leq T^{1/3}$. Exchanging the order of summation, it follows from (43) that

$$\begin{aligned} RV^2 &\ll RG \log^2 T + \sqrt{2} G \log T \sum_{n \leq TG^{-2} \log T} (-1)^n d(n) n^{-1/2} \times \\ &\times \sum_{s=1}^{S} \left(\left(\frac{t_s}{2\pi n} + \frac{1}{4} \right) - \frac{1}{2} \right)^{-1/2} \exp(-G^2 \cdots) \sin f(t_s, n) \\ &\ll RG \log^2 T + G \left(\sum_{n \leq TG^{-2} \log T} d^2(n) n^{-1} \right)^{1/2} \left(\sum_{n \leq TG^{-2} \log T} \left| \sum_{s=1}^{S} \cdots \right|^2 \right)^{1/2}, \end{aligned} \tag{44}$$

by the Cauchy-Schwarz inequality. We further have

$$\sum_{n \leq TG^{-2} \log T} \left| \sum_{s=1}^{S} \cdots \right|^2 = \sum_{s_1, s_2 \leq S} S_0,$$

where

$$\begin{aligned} S_0 := &\sum_{n \leq TG^{-2} \log T} \left(\left(\frac{t_{s_1}}{2\pi n} + \frac{1}{4} \right)^{1/2} - \frac{1}{2} \right)^{-1/2} \\ &\times \exp\left(-G^2 \left(\operatorname{arsinh} \sqrt{\pi n / (2t_{s_1})} \right)^2 \right) \left(\left(\frac{t_{s_2}}{2\pi n} + \frac{1}{4} \right)^{1/2} - \frac{1}{2} \right)^{-1/2} \\ &\times \exp\left(-G^2 \left(\operatorname{arsinh} \sqrt{\pi n / (2t_{s_2})} \right)^2 \right) \exp(if(t_{s_1}, n) - if(t_{s_2}, n)). \end{aligned}$$

Removing by partial summation monotonic coefficients from S_0, we are led to the crucial exponential sum

$$S_1 := \sum_{n \leq M} \exp(if(t_{s_1}, n) - if(t_{s_2}, n)) \qquad (M \leq TG^{-2} \log T). \tag{45}$$

The quality of the estimation of S_1 is limited by the scope of the present-day exponential sum techniques (see e.g., M.N. Huxley [4]). The terms $s_1 = s_2$ in (45) will eventually give rise to $R \ll_\varepsilon T^{1+\varepsilon}V^{-6}$, namely to a weak form of the sixth moment (25), but it does not seem likely that (25) can be reached (unconditionally) in this fashion. Observing that

$$\frac{\partial f(x,k)}{\partial x} = 2\,\mathrm{arsinh}\sqrt{\frac{\pi k}{2x}} \sim \sqrt{\frac{2\pi k}{x}} \quad (x \asymp T, k \le TG^{-2}\log T), \tag{46}$$

setting

$$f(u) := f(t_{s_1}, u) - f(t_{s_2}, u), \quad F := |t_{s_1} - t_{s_2}|(KT)^{-1/2},$$

we have (see [8, Chapters 1-2] for the relevant exponent pair technique) that S_1 may be split into $O(\log T)$ subsums of the type

$$\sum_{K<k\le K'\le 2K} \exp(if(k)) \ll F^\kappa K^\lambda + F^{-1}$$
$$\ll J^\kappa T^{-\kappa/2}K^{\lambda-\kappa/2} + (KT)^{1/2}|t_{s_1} - t_{s_2}|^{-1} \quad (K \le TG^{-2}\log T),$$

provided that $|t_{s_1} - t_{s_2}| \le J(\ll T)$, and (κ, λ) is a (one-dimensional) exponent pair. Choosing $(\kappa, \lambda) = (\frac{1}{2}, \frac{1}{2})$, $J = T^{-\varepsilon}G^3$ we obtain (17) (with T^ε in place of $\log^{17} T$). Namely with $J = T^{-\varepsilon}G^3$ the number of points $R = R_0$ to be estimated satisfies $R_0 \ll_\varepsilon T^{1+\varepsilon}G^{-3}$, hence dividing $[T/2, 5T/2]$ into subintervals of length not exceeding J one obtains

$$R \ll R_0(1 + T/J) \ll_\varepsilon T^{2+\varepsilon}G^{-6} \ll_\varepsilon T^{2+\varepsilon}V^{-12},$$

which easily yields (17) (with T^ε in place of $\log^{17} T$). This analysis was carried in detail in [8, Chapter 8], where the possibilities of choosing other exponent pairs besides $(\kappa, \lambda) = (\frac{1}{2}, \frac{1}{2})$ were discussed.

Another type of a similar estimate was obtained by the author in [15] (for the analysis of sums of moments over well-spaced points $\{t_r\} \in [T, 2T]$ the reader is referred to [10]). This result will be stated here as

Theorem 3. *Let $T \le t_1 < \cdots < t_R \le 2T$ be points such that $|\zeta(\frac{1}{2} + it_r)| \ge VT^{-\varepsilon}$ with $t_{r+1} - t_r \ge V \ge T^{\frac{1}{10}+\varepsilon}$ for $r = 1, \ldots, R-1$. Then, for any fixed integer $M \ge 1$,*

$$R \ll_\varepsilon T^{\varepsilon - M/2}V^{-2} \max_{K \le T^{1+\varepsilon}V^{-4}} \times$$
$$\times \int_{T/2}^{5T/2} \varphi(t)\Big|\sum_{K\le k\le K'\le 2K} (-1)^k d(k)k^{-1/4}\exp(2i\sqrt{2\pi kt} + cik^{3/2}t^{-1/2})\Big|^{2M} \mathrm{d}t, \tag{47}$$

where $c = \sqrt{2\pi^3}/6$ and $\varphi(t)$ is a non-negative, smooth function supported in $[T/2, 5T/2]$ such that $\varphi(t) = 1$ for $T \le t \le 2T$.

The case $M = 1$ quickly leads to a weakened form of the fourth moment estimate, namely $\int_0^T |\zeta(\frac{1}{2} + it)|^4\,\mathrm{d}t \ll_\varepsilon T^{1+\varepsilon}$. The case $M = 2$ of (46), by the use of Lemma 5 and Lemma 6, will lead again to a weakened form of the twelfth moment bound (17) (with $T^{2+\varepsilon}$).

Finally we present a new result, which connects bounds for moments of $|\zeta(\frac{1}{2} + it)|$ to bounds of moments of $J_k(t, G)$. This is

Theorem 4. *Suppose that*

$$\int_T^{2T} J_k^m(t,G)\,\mathrm{d}t \ll_\varepsilon T^{1+\varepsilon} \tag{48}$$

holds for some fixed $k, m \in \mathbb{N}$ *and* $G \geq T^{\alpha_{k,m}+\varepsilon}$, $0 \leq \alpha_{k,m} < 1$. *Then*

$$\int_0^T |\zeta(\tfrac{1}{2}+it)|^{2km}\,\mathrm{d}t \ll_\varepsilon T^{1+(m-1)\alpha_{k,m}+\varepsilon}. \tag{49}$$

Proof of Theorem 4: Set $L_k(t,G) = \int_{t-G}^{t+G} |\zeta(\frac{1}{2}+iu)|^{2k}\,\mathrm{d}u$. Note that if $\mu(\cdot)$ denotes measure, the bound

$$\mu\Big(t \in [T,2T] : L_k(t,G) \geq GU\Big) \ll_\varepsilon T^{1+\varepsilon}U^{-m} \qquad (U>0) \tag{50}$$

follows from (7) and (48). We fix $G = G(T)$ and divide the sum over s in (43) into $O(\log T)$ subsums where $GU < L_k(\tau_s, G) \leq 2GU$. Then, for $U_0 (\gg 1)$ to be determined later, we have

$$\begin{aligned}\sum_{s=1}^{S} L_k(\tau_s,G) &\ll GS\,U_0 + \log T \max_{U\geq U_0} \sum_{s,GU<L_k(\tau_s,G)\leq 2GU} L_k(\tau_s,G)\\ &\ll GS\,U_0 + GU\log T \max_{U\geq U_0} \sum_{s,GU<L_k(\tau_s,G)\leq 2GU} 1\\ &\ll_\varepsilon GS\,U_0 + \log T \max_{U\geq U_0} T^{1+\varepsilon}U^{1-m}\\ &\ll_\varepsilon GS\,U_0 + T^{1+\varepsilon}U_0^{1-m},\end{aligned}$$

since $m \geq 1$ and

$$\sum_{s,L_k(\tau_s,G)>GU} 1 \ll_\varepsilon T^{1+\varepsilon}U^{-m}G^{-1}. \tag{51}$$

Namely if $L_k(\tau_s,G) > GU$, then

$$L_k(t,2G) \geq L_k(\tau_s,G) > GU \qquad (\text{for } |t-\tau_s| \leq G).$$

As we can split the sequence of points $\{\tau_s\}$ into five subsequences, say $\{\tau_s'\}$, such that $|\tau_{s_1}' - \tau_{s_2}'| \geq 5G$ for $s_1 \neq s_2$, we see that

$$G \sum_{s,L_k(\tau_s,G)>GU} 1 \ll \mu\Big(t \in [T,\,2T] : L_k(t,2G) \geq GU\Big),$$

and (51) follows from (50). The choice

$$U_0 = \left(\frac{T}{SG}\right)^{1/m} \qquad (\gg 1)$$

yields

$$\sum_{s=1}^{S} L_k(\tau_s,G) \ll_\varepsilon T^{1/m+\varepsilon}S^{1-1/m}G^{1-1/m}. \tag{52}$$

By (43) we have $RV^{2k} \ll \log T \sum_{s=1}^{S} L_k(\tau_s, G)$, hence from (52) we obtain

$$R \ll_\varepsilon T^{1+\varepsilon} G^{m-1} V^{-2km}, \tag{53}$$

since $S \le R$. Thus (49) easily follows from (53), on taking $G = T^{\alpha_{k,m}+\varepsilon}$.

This completes the proof of Theorem 4. The values $\alpha_{1,2} = 0$ (Theorem 1) and $\alpha_{2,2} = \frac{1}{2}$ (Theorem 2) yield, respectively,

$$\int_0^T |\zeta(\tfrac{1}{2} + it)|^4 \,\mathrm{d}t \ll_\varepsilon T^{1+\varepsilon}, \qquad \int_0^T |\zeta(\tfrac{1}{2} + it)|^8 \,\mathrm{d}t \ll_\varepsilon T^{3/2+\varepsilon}. \tag{54}$$

The bounds in (54) are, of course, well-known, but they are (up to the factor T^ε) the sharpest known ones, and the bound for the fourth moment is essentially of the correct order of magnitude. Other values of $\alpha_{k,m}$ $(k = 1, 2)$, furnished by Theorem 1 and Theorem 2, do not yield any new bounds, as can be readily checked. However, it seems that this approach is of interest, especially in view of recent results on the distribution of sums and differences of square roots of integers (cf. Lemma 4 and Lemma 5). □

References

[1] F.V. Atkinson, The mean value of the Riemann zeta-function, *Acta Math.* **81**(1949), 353–376.

[2] J.B. Conrey, D.W. Farmer, J.P. Keating, M.O. Rubinstein and N.C. Snaith, Integral moments of L-functions, *Proc. London Math. Soc.* (3) **91**(2005), 33–104.

[3] J.L. Hafner and A. Ivić, On the mean square of the Riemann zeta-function on the critical line. *J. Number Theory*, **32**(1989), 151–191.

[4] D.R. Heath-Brown, The twelfth power moment of the Riemann zeta-function, *Quart. J. Math.* (Oxford) **29**(1978), 443–462,

[5] D.R. Heath-Brown, The fourth moment of the Riemann zeta-function, *Proc. London Math. Soc.* (3)**38**(1979), 385–422.

[6] M.N. Huxley, *Area, Lattice Points and Exponential Sums*, Oxford Science Publications, Clarendon Press, Oxford, 1996

[7] A.E. Ingham, Mean-value theorems in the theory of the Riemann zeta-function, *Proc. London Math. Soc.* (2)**27**(1926), 273–300.

[8] A. Ivić, *The Riemann zeta-function*, John Wiley & Sons, New York, 1985.

[9] A. Ivić, *The mean values of the Riemann zeta-function*, LNs **82**, Tata Inst. of Fundamental Research, Bombay (distr. by Springer Verlag, Berlin etc.), 1991.

[10] A. Ivić, Power moments of the Riemann zeta-function over short intervals, *Arch. Mat.* **62** (1994), 418–424.

[11] A. Ivić, On the fourth moment of the Riemann zeta-function, *Publs. Inst. Math. (Belgrade)* **57(71)**(1995), 101–110.

[12] A. Ivić, On the ternary additive divisor problem and the sixth moment of the zeta-function, *Sieve Methods, Exponential Sums, and their Applications in Number Theory* (eds. G.R.H. Greaves, G. Harman, M.N. Huxley), Cambridge University Press, Cambridge, 1996, 205–243.

[13] A. Ivić, On some results concerning the Riemann Hypothesis, in *Analytic Number Theory*, LMS LNS 247, Cambridge University Press, Cambridge, 1997, pp. 139–167.

[14] A. Ivić, On sums of Hecke series in short intervals, *J. de Théorie des Nombres Bordeaux* **13**(2001), 453–468.

[15] A. Ivić, Sums of squares of $|\zeta(\frac{1}{2} + it)|$ over short intervals, Max-Planck-Institut für Mathematik, Preprint Series 2002(**52**), 12 pp. arXiv:math.NT/0311516,

`http://front.math.ucdavis.edu/mat.NT/0311516`.

[16] A. Ivić and Y. Motohashi, The mean square of the error term for the fourth moment of the zeta-function, *Proc. London Math. Soc.* (3)**66**(1994), 309–329.

[17] A. Ivić and Y. Motohashi, The fourth moment of the Riemann zeta-function, *J. Number Theory* **51**(1995), 16–45.

[18] A. Ivić and P. Sargos, On the higher moments of the error term in the divisor problem, *Illinois J. Math.*, to appear.

[19] M. Jutila, Riemann's zeta-function and the divisor problem, *Arkiv Mat.* **21**(1983), 75–96 and II, ibid. **31**(1993), 61–70.

[20] K. Matsumoto, Recent developments in the mean square theory of the Riemann zeta and other zeta-functions, in *Number Theory*, Birkhäuser, Basel, 2000, 241–286.

[21] Y. Motohashi, A relation between the Riemann zeta-function and the hyperbolic Laplacian, *Ann. Sc. Norm. Sup. Pisa, Cl. Sci.* IV ser. **22**(1995), 299–313.

[22] Y. Motohashi, *Spectral theory of the Riemann zeta-function*, Cambridge University Press, Cambridge, 1997.

[23] K. Ramachandra, A simple proof of the mean fourth power estimate for $L(\frac{1}{2} + it, \chi)$, *Ann. Sc. Norm. Sup. Pisa, Cl. Sci.* **1**(1974), 81–97.

[24] K. Ramachandra, *On the mean-value and omega-theorems for the Riemann zeta-function*, Tata Institute of Fundamental Research, Bombay, distr. by Springer Verlag, 1995.

[25] O. Robert and P. Sargos, Three-dimensional exponential sums with monomials, *J. reine angew. Math.* **591**(2006), 1–20.

[26] E.C. Titchmarsh, *The theory of the Riemann zeta-function* (2nd ed.), University Press, Oxford, 1986.

[27] K.-M. Tsang, Higher power moments of $\Delta(x)$, $E(t)$ and $P(x)$, *Proc. London Math. Soc.* (3) **65**(1992), 65–84.

Aleksandar Ivić
Katedra Matematike RGF-a
Universiteta u Beogradu,
Du sina 7, 11000 Beograd,
Serbia (Yugoslavia).

email: ivic@rgf.bg.ac.yu,
aivic@matf.bg.ac.yu

The Riemann Zeta Function and Related Themes – 2006, pp. 99–114

On product partitions and asymptotic formulas

I. Kátai[*†] and M.V. Subbarao[†‡]

Dedicated to Professor K. Ramachandra on his 70th birthday

Abstract

Let $p^*(n)$ denote the number of product partitions of an integer $n > 1$ and $q^*(n)$ the number of product partitions of n into distinct parts, and $p^*(1) = q^*(1) = 1$. A. Oppenheim and, independently and in a more detailed manner, G. Szekeres and P. Turan obtained an asymptotic result for the average order of $p^*(n)$.

We here prove a general asymptotic result for the average value of a function $f(n)$ defined for $n > 1$ by

$$\Pi\,(1 + e(n)n^{-s}) = 1 + \sum f(n)n^{-s},$$

(products and summations being for all integers $n \geq 2$), where $e(n)$ is subject to some specified restrictions. The class C of such functions $f(n)$ includes, besides many others, the function $q^*(n)$, for which no result exists in the current literature regarding the asymptotics of its summatory function $\sum_{n\leq x} q^*(n)$.

1 Introduction and preliminaries

The study of product partitions began with a paper by P.A. MacMahon [6] in 1924 which was, almost immediately followed by two papers by A. Oppenheim ([7], [8]). We recall that by a product partition of an integer $n > 1$, we mean an unordered representation of n as a product of integers $m_1, m_2, \ldots$ each greater than unity, the integer $m_1, \ldots$ being called the 'parts' of the product partition. We use the notation $p^*(n)$ (respectively, $q^*(n)$) for the number of unrestricted product partitions of n (respectively, product partitions of n into distinct parts. We define $p^*(1) = q^*(1) = 1$.

Recently, another important product partition function $e^*(n)$ was introduced by Subbarao and Verma [9], $e^*(n)$ is defined as the excess of the number of product partitions of n into an even number of distinct parts over those into an odd number of such parts. Thus for example, we have $p^*(12) = 4$, $q^*(12) = 3$ and $e^*(12) = 1$. As is easily seen, we have, for Re $s > 1$,

$$\prod_{n=2}^{\infty}\left(1 - \frac{1}{n^s}\right)^{-1} = 1 + \sum_{n=2}^{\infty} p^*(n)n^{-s}$$

$$\prod_{n=2}^{\infty}\left(1 + \frac{1}{n^s}\right) = 1 + \sum_{n\geq 2}^{\infty} q^*(n)n^{-s}$$

$$\prod_{n=2}^{\infty}\left(1 - \frac{1}{n^s}\right)^{-1} = 1 + \sum_{n=2}^{\infty} e^*(n)n^{-s}.$$

[*]Financially supported by the second author's NSERSC Discovery Grant.

[†]Financial supported by the Applied Number Theory Group of HAS and by the Hungarian Scientific Research Fund.

[‡]Editors regret to record the passing away of Prof. M. V. Subbarao on February 15, 2006.

We already know some results concerning the asymptotic behaviour or maximal order of these functions. Thus, correcting an earlier result of A. Oppenheim [7], E.R. Canfield, P. Erdös and C. Pomerance [3] proved that max $(p^*(n) : n \le x) = x/ (L(x))^{1+0(1)}$, where

$$L(x) = \exp\,(x_1 x_3/x_2).$$

For a more detailed account of this result and results for lower bound for $q^*(n)$ we refer to their paper [3].

If $h(k)$ denotes the value of $q^*(n)$ when n is the product of any k distinct primes, then de Bruijn [4] proved that as $k \to \infty$,

$$\log h(k) = k\,(k_1 - k_2 - 1 + (k_2 + 1))\,/k_1 + k_2^2/2k_1^2 + O(k_2^3/k_1^3).$$

If $g(k)$ denotes the value of $e^*(n)$ when n is the product of any k distinct primes, Subbarao and Verma [9] proved that

$$\limsup_{k\to\infty} \frac{\log\,|g(k)|}{k\,\log\,k} = 1.$$

Here, $k_1 = \log k$, $k_2 = \log k_1, \ldots$.

In this paper, we shall prove using some of the ideas from [8] and [11], a general theorem on the average order of a class C of functions $f(n)$ defined by the generating function

$$F(s) := \sum_{n=1}^{\infty} \frac{f(n)}{n^s} = \prod_{n=2}^{\infty} \left(1 + \frac{e(n)}{n^s}\right),$$

where $e(n)$ is a non-negative function subject to some properties specified in the next section. It may be stated here that the class C includes the functions $q^*(n)$ among others.

Note that $f(n)$ may be interpreted as a weighted product partition function in the following sense. To each product partition $n = m_1 m_2 \ldots m_r$ of n into distinct parts, attach a weight $e(m_1)e(m_2)\ldots e(m_r)$. Then $f(n) = \sum e(1)e(2)\ldots e(r)$, summation being over all the product partitions of n into distinct parts.

We have also obtained results on an asymptotic formula for $\sum_{n\le x} e^*(n)$, with an 0-term estimate for the error term, but the results are not included here.

Notation

Throughout we follow the following notation:

(i) N = set of positive integers

(ii) $p, p_1, p_2, \ldots$ represent primes, unless specifically stated otherwise

(iii) $d(n)$ = the number of positive divisors of n

(iv) $d_k(n) = \sum_{a_1 a_2 \ldots a_k = n} 1$, $a_i > 0$ integer, $i = 1, \ldots, k$

(v) $[x]$ = largest integer $\le x$

(vi) $\zeta(s)$ is the Riemann zeta function

(vii) For any positive number x (variable or fixed) $x_1 = \log x$, $x_2 = \log x_1, \ldots$.

(viii) $c, c_1, \ldots$ denote absolute constants

2 The functions $f(n)$ and the associated integral

As already stated, our purpose here is to establish a general asymptotic formula for the summatory function generated by the Dirichlet series in the expansion of $\Pi(1 + e(n)n^{-s})$ when $e(n)$ is suitably restricted. For this purpose, we use among our devices, some of the ideas of Oppenheim, and Szekeres-Turan in their proof of

$$\sum_{n\le x} p^*(n) \sim \frac{e^{2\sqrt{\log x}}}{2\sqrt{\pi}\,(\log x)^{3/4}}. \tag{1}$$

2.1 The functions $e(n)$, $f(n)$ and the integral for $D(x)$

Let $e(n) \ge 0$ $(n = 2, 3, \dots)$ be an arithmetic function subject to the assumption: $e(n) < n$, $e(n) = O(n^{\varepsilon})$ for each $\varepsilon > 0$. Further, set

$$U(s) = \sum_{n=2}^{\infty} \frac{e(n)}{n^s}$$

and assume that

$$U(s) = \frac{A}{(s-1)^{\beta}} + u_1(s), \quad (s = \sigma + it)$$

in the half plane Re $s \ge 1$ where A, β or are positive constants and $u_1(s)$ is bounded in $\sigma \ge 1$, $|t| \le 1$.

Assumptions: *Let* $e(n) \ge 0$ $(n = 2, 3, \dots)$, $e(n) = O(n^{\varepsilon})$ *for each* $\varepsilon > 0$, $|e(n)| < n$ $(n = 2, 3, \dots)$. *Let*

$$U(s) = \sum_{n=2}^{\infty} \frac{e(n)}{n^s}. \tag{2}$$

Assume that

$$U(s) = \frac{A}{(s-1)^{\beta}} + u_1(s) \quad (s = \sigma + it) \tag{3}$$

in the half plane Re $s \ge 1$, *where* A, β *are positive constants* $u_1(s)$ *is bounded in* $\sigma \ge 1$, $|t| \le 1$. *Assume furthermore that*

$$|U(1 + it)| < (1 - \varepsilon)\log 2|t| \quad \textit{if} \quad |t| \ge 1. \tag{4}$$

Here $\varepsilon > 0$ *is an arbitrary positive constant.*

Let

$$F(s) := \sum_{n=1}^{\infty} \frac{f(n)}{n^s} = \prod_{n=2}^{\infty} \left(1 + \frac{e(n)}{n^s}\right). \tag{5}$$

Since

$$\log F(s) = \sum_{k=1}^{\infty} \frac{(-1)^{k-1}}{k} \sum_{n=2}^{\infty} \frac{e(n)}{n^{ks}} = \sum_{k=1}^{\infty} \frac{(-1)^{k-1}}{k} U(ks),$$

we obtain that

$$F(\sigma + it) = \exp\left(\frac{A}{(s-1)^{\beta}} + u_2(s)\right) \quad \text{in} \quad \sigma \ge 1,\ |t| \le 1, \tag{6}$$

where $u_2(s)$ is bounded here, furthermore

$$\begin{aligned}|F(\sigma+it)| &\le \exp\left(\operatorname{Re} U(\sigma+it)+c_1\right)\\ &\le \exp\left((1-\varepsilon)\log 2|t|+c_1\right)\\ &\le e^{c_1}(2|t|)^{1-\varepsilon}\end{aligned} \tag{7}$$

in $\sigma \ge 1$, $|t| \ge 1$.

Let

$$\mathcal{D}(x) := \sum_{n\le x} f(n)\left(1-\frac{n}{x}\right). \tag{8}$$

We shall start from the well known formula

$$\mathcal{D}(x) = \frac{1}{2\pi i}\int_{2-i\infty}^{2+i\infty} F(s)\,\frac{x^s}{s(s+1)}\,ds. \tag{9}$$

We can move the integration line of (9) into L, where L consists of the intervals I_1, I_2, I_3, I_4 and that of the half-circle C defined exactly as follows:

$$I_1 = \{1+it \mid t\in[1,\infty)\}, \qquad I_2 = \{1+it \mid t\in[r,1)\}$$

$$I_3 = \{1-it \mid t\in[1,\infty)\}, \qquad I_4 = \{1-it \mid t\in[r,1)\}$$

$$C = \left\{s = 1+re^{i\theta},\quad -\tfrac{\pi}{2}\le\theta\le\tfrac{\pi}{2}\right\}.$$

Here

$$r = \left(\frac{\beta A}{x_1}\right)^{1/\beta+1}. \tag{10}$$

Let furthermore

$$\alpha = \frac{\beta+1}{\beta}, \quad B := rx_1 = \frac{\beta A}{r^\beta}, \tag{11}$$

$$\mathcal{D} = \frac{\beta+1}{2}\,B. \tag{12}$$

From (7) we obtain that

$$\left|\frac{1}{2\pi i}\int_{I_j} F(s)\,\frac{x^s}{s(s+1)}\,ds\right| = O(x), \text{ if } \quad j=1,4. \tag{13}$$

On the intervals I_2, I_3

$$|F(1+it)| \le O(1)\exp\left(\frac{A}{r^\beta}\cos\beta\pi/2\right),$$

since

$$\frac{1}{\beta}\cos\beta\pi/2 \le \frac{1}{\beta} = \alpha-1,$$

we obtain that

$$\left|\frac{1}{2\pi i}\int_{I_j} F(s)\,\frac{x^s}{s(s+1)}\,ds\right| \ll x\exp(-B)\exp(B\alpha). \tag{14}$$

Thus

$$\sum_{j=1}^{4}\left|\frac{1}{2\pi i}\int_{I_j} F(s)\,\frac{x^s}{s(s+1)}\,ds\right| \ll x\,\exp\left(B(\alpha-1)\right). \tag{15}$$

Now we consider the integral over C.

From (6),

$$\frac{F(s)}{s(s+1)} = \exp\left(\frac{A}{(s-1)^\beta}\right)\frac{\exp\left(u_2(s)\right)}{s(s+1)}. \tag{16}$$

Furthermore,

$$\frac{\exp\left(u_2(s)\right)}{s(s+1)} = a_0 + a_1(s-1) + \cdots + a_R(s-1)^R + (s-1)^{R+1}g(s-1) \tag{17}$$

where R is an arbitrary fixed integer, $a_0, a_1, \ldots, a_R$ are suitable constants, $a_0 \neq 0$, and $g(s-1)$ is bounded in the disc $|s-1| \leq r$.

Let

$$\mathcal{S}(h) = \frac{1}{2\pi i}\int_C \exp\left(\frac{A}{(s-1)^\beta} + (s-1)x_1\right)\cdot(s-1)^{h-1}ds, \tag{18}$$

$$\mathcal{T} = \frac{1}{2\pi i}\int_C \exp\left(\frac{A}{(s-1)^\beta} + (s-1)x_1\right)\cdot(s-1)^{R+1}g(s-1)ds. \tag{19}$$

From (16), (17) we obtain that

$$\frac{1}{2\pi i x}\int_C F(s)\,\frac{x^s}{s(s+1)}\,ds = \sum_{h=1}^{R+1} a_{h-1}\mathcal{I}(h) + \mathcal{T}. \tag{20}$$

Let us introduce the following notations:

$$p(\theta) = \frac{\cos\beta\theta}{\beta} + \cos\theta, \quad q(\theta) = -\frac{\sin\beta\theta}{\beta} + \sin\theta, \tag{21}$$

$$u(\theta) = Bp(\theta), \quad v_h(\theta) = Bq(\theta) + h\theta. \tag{22}$$

By the substitution $s-1 = re^{i\theta}$, $ds = ire^{i\theta}d\theta$, we have

$$\mathcal{I}(h) = \frac{r^h}{2\pi}\int_{-\pi/2}^{\pi 2} \exp\left(Bp(\theta) + iv_h(\theta)\right)d\theta, \tag{23}$$

$$|\mathcal{T}| \leq \frac{r^{R+2}}{2\pi}\int_{-\pi/2}^{\pi/2} \exp\left(Bp(\theta)\right)d\theta. \tag{24}$$

Since $u(\theta) = u(-\theta)$, $v_h(-\theta) = -v_h(\theta)$, therefore

$$\mathcal{I}(h) = \frac{r^h}{\pi}\,S(h), \tag{25}$$

where

$$S(h) = \int_0^{\pi/2} \exp\left(u(\theta)\right)\,\cos\,v_h(\theta)d\theta. \tag{26}$$

We shall estimate $S(h)$ as well as the integral

$$S^* = \int_0^{\pi/2} \exp\,(u(\theta))\,d\theta. \tag{27}$$

Let σ be a fixed constant, $0 < \sigma < \frac{1}{2\alpha}$, and let

$$(\Delta_x =)\Delta = x_1^{-\frac{1}{3}(\frac{1}{\alpha}+\sigma)}. \tag{28}$$

Then $\Delta^3 B \to 0$, $\Delta^2 B \to \infty$ as $x \to \infty$. We shall write

$$S(h) = S_1(h) + S_2(h) = \int_0^{\Delta} + \int_{\Delta}^{\pi/2} \tag{29}$$

$$S^* = S_1^* + S_2^* = \int_0^{\Delta} + \int_{\Delta}^{\pi/2}. \tag{30}$$

Since

$$\max_{\Delta \le \theta \le \pi/2} p(\theta) \le \alpha - (\beta + 1)\Delta^2,$$

therefore

$$S_2(h) \ll \exp(\alpha B)\exp\left(-B(\beta+1)\Delta^2\right) \tag{31}$$

$$S^* \ll \exp(\alpha B)\exp\left(-B(\beta+1)\Delta^2\right). \tag{32}$$

The main problem is to compute the asymptotic expansion of $S_1(h)$, and of S_1^*. Before doing this we shall first prove a lemma and recall the notion of dominating power series, since it plays a useful part in our subsequent discussions.

3 Lemma 1 and a note on dominating series

3A. Lemma 1

Lemma 1. *Let $p \ge 0$ be a real number. then*

$$\int_0^{\infty} \exp(-\mathcal{D}\theta^2)\theta^p\,d\theta = \frac{\Gamma\left(\frac{p+1}{2}\right)}{2\mathcal{D}^{(p+1)/2}}, \tag{33}$$

where Γ is Euler's gamma function, $\mathcal{D}$ is defined in (12).

Let

$$M_p : \int_0^{\Delta} \exp(-\mathcal{D}\theta^2)\theta^p\,d\theta, \tag{34}$$

where $\Delta = \Delta_x$ is defined in (28).

Then

$$M_p = \frac{\Gamma\left(\frac{p+1}{2}\right)}{2\mathcal{D}^{(p+1)/2}} + O\left(\frac{\Delta^{p-1}}{\mathcal{D}}\right)\exp(-\mathcal{D}\Delta^2), \tag{35}$$

where the constant implied by the error term may depend on p.

Proof. By using the substitution $\eta = \mathcal{D}\theta^2$ in (34), we deduce that

$$M_p = \frac{1}{2\mathcal{D}^{\frac{p+1}{2}}} \int_0^{\mathcal{D}\Delta^2} \exp(-\eta)\eta^{\frac{p-1}{2}}\, d\eta \tag{36}$$

and similarly that the left hand side of (33)

$$= \frac{1}{2\mathcal{D}^{(p+1)/2}} \int_0^{\infty} \exp(-\eta)\eta^{\frac{p-1}{2}}\, d\eta\,,$$

which by the known identity

$$\int_0^{\infty} e^{-\eta} \cdot \eta^{s-1} d\eta = \Gamma(s)$$

gives (33). To prove (35), we start from

$$M_p = \frac{\Gamma\left(\frac{p+1}{2}\right)}{2\mathcal{D}^{(p+1)/2}} - \frac{1}{2\mathcal{D}^{(p+1)/2}} \int_{\mathcal{D}\Delta^2}^{\infty} \exp(-\eta)\eta^{\frac{p-1}{2}}\, d\eta.$$

By partial integration,

$$\int_{\mathcal{D}\Delta^2}^{\infty} \exp(-\eta)\eta^{\frac{p-1}{2}}\, d\eta = \int_{\mathcal{D}\Delta^2}^{\infty} (-\exp(-\eta))'\ \eta^{\frac{p-1}{2}}\, d\eta,$$

where ( )′ denotes differentiation of ( ) with respect to η,

$$= \exp(-\mathcal{D}\Delta^2)(\mathcal{D}\Delta^2)^{\frac{p-1}{2}} + \frac{p-1}{2} \cdot \int_{\mathcal{D}\Delta^2}^{\infty} \exp(-\eta) \cdot \eta^{\frac{p-3}{2}}\, d\eta.$$

Iterating this procedure, (35) easily follows. □

2B. Dominating power series

Let

$$\varphi(x_1, x_2, \ldots, x_k) = \sum C_{m_1,\ldots,m_k} x_1^{m_1} \ldots x_k^{m_k} \tag{37}$$

be a power series of the variables $x_1, \ldots, x_k$.

Let $|C_{m_1,\ldots,m_k}| \le d_{m_1,\ldots,m_k}$ and

$$\psi(x_1, x_2, \ldots, x_k) = \sum d_{m_1,\ldots,m_k} x_1^{m_1} \ldots x_k^{m_k}. \tag{38}$$

We then say that the right hand side of (38) dominates the power series defined in (37). We say also, that ψ dominates φ.

Let $\sigma_j(y_1, \ldots, y_h)$ be dominated by $\tau_j(y_1, \ldots, y_h)$ $(j = 1, \ldots, k)$, and $\varphi(x_1, x_2, \ldots, x_k)$ is dominated by $\psi(x_1, x_2, \ldots, x_k)$. Then clearly

$$\Lambda(y_1, \ldots, y_h) := \psi\left(\sigma_1(y_1, \ldots, y_h), \ldots, \sigma_k(y_1, \ldots, y_h)\right)$$

is dominated by

$$M(y_1, \ldots, y_h) := \varphi\left(\tau_n(y_1, \ldots, y_h), \ldots, \tau_k(y_1, \ldots, y_h)\right).$$

4 Lemmas 2 and 3

Let

$$\lambda(\theta) = \frac{1-\cos\beta\theta}{\beta} + (1-\cos\theta). \tag{39}$$

Then $p(\theta) = \alpha - \lambda(\theta)$, and

$$S_1(h) = \exp(\alpha B)\int_0^\Delta \exp(-B\lambda(\theta))\cos v_h(\theta)d\theta, \tag{40}$$

$$S_1^* = \exp(\alpha B)\int_0^\Delta \exp(-B\lambda(\theta))\,d\theta. \tag{41}$$

The power series expansion of $\lambda(\theta)$ is

$$\lambda(\theta) = \frac{\beta+1}{2!}\theta^2 - \frac{\beta^3+1}{4!}\theta^4 + \frac{\beta^5+1}{6!}\theta^6 - \cdots . \tag{42}$$

Let $\lambda_1(\theta) = \frac{\beta^3+1}{4!}\theta^4 - \frac{\beta^6+1}{6!}\theta^6 + \cdots$, i.e. $\lambda(\theta) = \frac{\beta+1}{2}\theta^2 - \lambda_1(\theta)$.

Thus

$$\exp(-B\lambda(\theta)) = \exp(-\mathcal{D}\theta^2)\exp(B\lambda_1(\theta)),$$

$$\exp(-B\lambda(\theta))\cos v_h(\theta) = \exp(-\mathcal{D}\theta^2)\exp(B\lambda_1(\theta))\cos v_h(\theta).$$

Let

$$\exp(B\lambda_1(\theta)) = \sum b_{u,v}B^u\theta^v, \tag{43}$$

$$\exp(B\lambda_1(\theta))\cos v_h(\theta) = \sum C_{u,v}^{(h)}B^u\theta^v. \tag{44}$$

Let

$$\lambda_2(\theta) = \frac{\beta^3+1}{4!}\theta^4 + \frac{(\beta^5+1)}{6!}\theta^6 + \cdots .$$

Then the function $f(B,\theta) := \exp(B\lambda_2(\theta))$ dominates $\exp(B\lambda_1(\theta))$,

$$f(B,\theta) = \sum d_{u,v}B^u\cdot\theta^v$$

and the power series is convergent for every B, θ, consequently

$$\sum d_{u,v} < \infty.$$

Since

$$q(\theta) := \sum_{j=1}^{\infty} \frac{(-1)^j}{(2j+1)!}(1-\beta^{2j})\theta^{2j+1},$$

therefore $q(\theta)$ is dominated by $\exp((\beta+1)\theta)$, and $Bq(\theta) + h\theta$ is dominated by $B\exp((\beta+1)\theta) + h\theta$, and since $\cos y$ is dominated by e^y, we obtain that $\cos v_h(\theta)$ is dominated by $\exp\{B\exp((\beta+1)\theta) + h\theta\}$, therefore (44) is dominated by

$$f(B,\theta)\cdot\exp\{B\exp((\beta+1)\theta) + h\theta\} =: f_h(B,\theta).$$

Let

$$f_h(B,\theta) = \sum d_{u,v}^{(h)}B^u\theta^v.$$

Since the power series is everywhere convergent, therefore

$$\sum d_{u,v}^{(h)} < \infty.$$

Let us observe furthermore that $b_{u,v} = 0$ if $v \le 4u - 1$, since $\lambda_1(\theta) = \frac{\beta^3+1}{4!}\,\theta^4 - \cdots$. Similarly, since $q(\theta) = \frac{\beta^2-1}{3!}\,\theta^3 + \cdots$, we obtain that $C_{u,v}^{(h)} = 0$ if $v \le 3u - 1$.

We discuss first the asymptotic expansion of

$$\int_0^{\Delta} \exp\,(-B\lambda_1(\theta))\,d\theta.$$

Let R_1 and R_2 be arbitrary positive integers such that $R_2 \ge 4R_1$.

We have

$$\begin{aligned}\left|\exp\,(B\lambda_1(\theta)) - \sum_{u=0}^{R_1}\sum_{v=4u}^{R_2} b_{u,v}B^u\cdot\theta^v\right| &\le \sum_{u>R_1}\sum_{v\ge 4u} d_{u,v}B^u\theta^v + \sum_{u\le R_1}\sum_{v>R_2} d_{u,v}\cdot B^u\cdot\theta^v \\ &= \sum_1 + \sum_2 .\end{aligned}$$

We can see easily that $\sum_1 \ll (B\theta^4)^{R_1+1}$, and $\sum_2 \ll B^{R_1}\theta^{R_2+1}$.

Thus, from Lemma 1, we obtain that

$$\left|\int_0^{\Delta} \exp(-B\lambda(\theta))\,d\theta - \sum_{u\le R_1}\sum_{v\le R_2} b_{u,v}\,\frac{\Gamma\left(\frac{v+1}{2}\right)}{2}\,\frac{B^u}{\mathcal{D}^{(v+1)/2}}\right| \ll \frac{B^{R_1}\Gamma\left(\frac{4R_1+3}{2}\right)}{\mathcal{D}^{(4R_1+3)/2}}\,. \tag{45}$$

Here we observed that in Lemma 1 the error term for the expression of M_p is less than $O(x_1^{-k})$, where k is an arbitrary large number.

We have

$$B^u\cdot\mathcal{D}^{-\frac{v+1}{2}} = \left(\frac{2}{\beta+1}\right)^{\frac{v+1}{2}} B^{u-\frac{v+1}{2}}\,.$$

From (45) we have

$$\begin{aligned}\int_0^{\Delta} \exp(-B\lambda(\theta))\,d\theta &= \sum_{u=0}^{R_1}\sum_{v=4u}^{R_2} \frac{1}{2}b_{u,v}\Gamma\left(\frac{v+1}{2}\right)\left(\frac{2}{\beta+1}\right)^{\frac{v+1}{2}} B^{u-\frac{v+1}{2}} \\ &\quad + O\left(B^{-R_1-\frac{3}{2}}\right).\end{aligned} \tag{46}$$

Hence the following Lemma 2 readily follows.

Lemma 2. *Let R_1 be an arbitrary positive integer. Then for S^* defined by* (27), *we have*

$$\begin{aligned}S^* &= \exp(\alpha B)\left\{\frac{\mathcal{D}_1}{B^{1/2}} + \frac{\mathcal{D}_2}{(B^{1/2})^2} + \cdots + \frac{\mathcal{D}_{2R_1}}{(B^{1/2})^{2R_1}}\right\} \\ &\quad + O\left(\exp(\alpha B)\cdot(B^{-1/2})^{(2R_1+1)}\right)\end{aligned} \tag{47}$$

where $\mathcal{D}_1, \mathcal{D}_2, \ldots, \mathcal{D}_{2R_1}$ are suitable constants.

As a byproduct we obtain that

$$|\mathcal{T}| = O\left(\frac{r^{R+2}}{\sqrt{B}} \exp(\alpha B)\right). \tag{48}$$

($\mathcal{T}$ is defined by (13).)

Proof of Lemma 2: (47) directly follows from (46) and from (32). (48) follows from (47), (32), (24).

Let R_1 be arbitrary, $R_2 \geq 3R_1$. Arguing as earlier,

$$\begin{aligned}\left|\exp(b\lambda_1(\theta))\cos v_h(\theta) - \sum_{u=0}^{R_1}\sum_{v=3u}^{R_2} b_{u,v}^{(h)} B^u \cdot \theta^v\right| &\leq \sum_{u>R_1}\sum_{v\geq 3u} d_{u,v}^{(h)} B^u \theta^v + \sum_{u\leq R_1}\sum_{v>R_2} d_{u,v}^{(h)} B^u \theta^v \\ &= \sum_1 + \sum_2 .\end{aligned}$$

One can see that

$$\sum_1 \ll (B\theta^3)^{R_1+1}, \quad \sum_2 \ll B^{R_1}\theta^{R_2+1}.$$

Thus, by Lemma 1,

$$\begin{aligned}&\left|\int_0^{\Delta} \exp(-B\lambda(\theta))\cos v_h(\theta) d\theta - \sum_{u=0}^{R_1}\sum_{v=3u}^{R_2} \frac{1}{2} b_{u,v}^{(h)} \Gamma\left(\frac{v+1}{2}\right)\left(\frac{2}{\beta+1}\right)^{\frac{v+1}{2}} B^{u-\frac{v+1}{2}}\right| \\ &\quad \ll B^{R_1+1}\cdot \mathcal{D}^{\frac{-3(R_1+1)-1}{2}} \ll B^{-\frac{R_1}{2}-2},\end{aligned} \tag{49}$$

whence we obtain the following.

Lemma 3. *Let R_h be an arbitrary positive integer, for every h in $[1, R+1]$. Then there exist suitable constants $E_1(h), \ldots, E_{R_h}(h)$ such that*

$$S(h) = \exp(\alpha B)\cdot \sum_{v=1}^{R_h} E_v(h)\cdot \frac{1}{(\sqrt{B})^v} + O\left(\exp(\alpha B)\cdot \frac{1}{(\sqrt{B})^{R_h+1}}\right) \tag{50}$$

Proof: (50) is a consequence of (49) and (31).

5 Theorem 1

From (9), (20), (48) we obtain that

$$\frac{\mathcal{D}(x)}{x} = \sum_{h=1}^{R+1} a_{h-1}\mathcal{I}(h) + O(\exp(B(\alpha-1))) + O\left(\exp(B\alpha)\cdot \frac{r^{R+2}}{\sqrt{B}}\right).$$

Then, from Lemma 3;

$$
\begin{aligned}
\frac{\mathcal{D}(x)}{x} &= \frac{1}{\pi}\sum_{h=1}^{R+1} a_{h-1} r^h \left\{\sum_{\nu=1}^{R_h} E_\nu(h)\cdot\frac{1}{(\sqrt{B})^\nu}\right\}\exp(\alpha B) \\
&\quad + O\left(\exp(B\alpha)\,\frac{r^{R+2}}{\sqrt{B}}\right) + O\left(\exp(B\alpha)\cdot\sum_{h=1}^{R+1}\frac{r^h}{(\sqrt{B})^{R_h+1}}\right).
\end{aligned}
\tag{51}
$$

Let

$$
\gamma := (\beta A)^{\frac{1}{\beta+1}}. \tag{52}
$$

From (10), (11) we obtain that

$$
r = \gamma\cdot x_1^{-1/(\beta+1)}, \qquad B = \gamma\cdot x_1^{1/\alpha}.
$$

Thus

$$
\begin{aligned}
\frac{r^h}{(\sqrt{B})^\nu} &= \gamma^{h-\frac{\nu}{2}}\cdot x_1^{-\frac{h}{\beta+1}-\frac{\nu}{2\alpha}} \\
&= \gamma^{h-\frac{\nu}{2}}\cdot\left(\frac{1}{x_1}\right)^{\frac{h+\frac{\nu}{2}\beta}{\beta+1}},
\end{aligned}
$$

$$
\frac{r^{R+2}}{\sqrt{B}} = O(1)\left(\frac{1}{x_1}\right)^{\frac{R+2+\frac{\beta}{2}}{\beta+1}}, \qquad \frac{r^h}{(\sqrt{B})^{R_h+1}} = O(1)\left(\frac{1}{x_1}\right)^{\frac{h+\frac{(R_h+1)}{2}\beta}{\beta+1}}.
$$

Let $\mathcal{E}$ be the set of real numbers $\eta_{h,\nu} := h + \frac{\nu}{2}\beta$, $(h = 1, 2, \ldots; \nu = 1, 2, \ldots)$.

Theorem 1. *Let R be an arbitrary positive integer. Let S_R be the set of those $\eta_{h,\nu}$ for which $\eta_{h,\nu} < R + 2 + \beta/2$ and $1 \le h \le R+1$ holds. Then, there exist constants $H(\eta_{h,\nu})$ $(\eta_{h,\nu} \in S_R)$ such that*

$$
\begin{aligned}
\frac{\mathcal{D}(x)}{x} &= \exp\left(\alpha\gamma\cdot x_1^{1/\alpha}\right)\sum_{\eta_{h,\nu}\in S_R} H(\eta_{h,\nu})\left(\frac{1}{x_1}\right)^{\frac{\eta_{h,\nu}}{\beta+1}} \\
&\quad + O\left(\exp(\alpha\gamma\cdot x_1^{1/\alpha})\cdot\left(\frac{1}{x_1}\right)^{\frac{R+2+\beta/2}{\beta+1}}\right)
\end{aligned}
\tag{53}
$$

Proof: The assertion is an immediate consequence of Lemma 2. We have to choose R_h to be so large that $\eta_{h,R_h} > R + 2 + \beta/2$, and apply (51).

Remark: More generally, a result as in the theorem can be proved under the more general conditions, namely that instead of (13), we assume that

$$
U(s) = V(s-1) + u_2(s),
$$

where

$$
V(s-1) = \frac{A_1}{(s-1)^{\beta_1}} + \frac{A_2}{(s-1)^{\beta_2}} + \cdots + \frac{A_r}{(s-1)^{\beta_r}},
$$

with $\beta_1 > \beta_2 > \cdots > \beta_r \ge 0$.

6 Theorem 2

Let

$$S(x) = \sum_{n \le x} f(n).$$

Then $S(x)$ is monotonically increasing, since $f(n) \ge 0$. We have

$$\int_0^x S(u)du = \sum_{n \le x} f(n)(x-n) = x\mathcal{D}(x).$$

Therefore

$$\begin{aligned} S(x) &\le \frac{1}{\delta x}\int_x^{s+\delta x} S(u)du \\ &= \frac{(x+\delta x)\mathcal{D}(x+\delta x) - x\mathcal{D}(x)}{\delta x} \\ &= \left(\frac{x}{\delta x}+1\right)(\mathcal{D}(x+\delta x)-\mathcal{D}(x)) + \mathcal{D}(x) \end{aligned} \tag{54}$$

and

$$S(x) \ge \frac{1}{\delta x}\int_{x-\delta x}^{x} S(u) = \left(\frac{x}{\delta x}-1\right)(\mathcal{D}(x)-\mathcal{D}(x-\delta x)) + \mathcal{D}(x). \tag{55}$$

Let

$$Q_{h,v}(x) = \exp(\alpha\gamma \cdot x_1^{1/\alpha}) \cdot x_1^{-\frac{\eta_{h_i,v}}{\beta+1}}. \tag{56}$$

From (53) we obtain that

$$\begin{aligned} \mathcal{D}(x) &= \sum_{\eta_{h,v}\in S_R} H(\eta_{h,v})xQ_{h,v}(x) + O(xQ_{R+2,1}(x)) \\ &= \mathcal{D}_1(x) + O(x \cdot Q_{R+2,1}(x)). \end{aligned} \tag{57}$$

From (54), (55) we deduce that

$$\begin{aligned} &\left|S(x) - \mathcal{D}(x) - x\frac{\mathcal{D}(x+\delta x)-\mathcal{D}(x)}{\delta x}\right| \\ &\le \max\left\{(\mathcal{D}(x+\delta x)-\mathcal{D}(x)), (\mathcal{D}(x)-\mathcal{D}(x-\delta x)) \right. \\ &\qquad \left. + \frac{\mathcal{D}(x+\delta x) - 2\mathcal{D}(x) + \mathcal{D}(x-\delta x)}{\delta x}\right\}. \end{aligned} \tag{58}$$

We have

$$\frac{\mathcal{D}(x+\delta x)-\mathcal{D}(x)}{\delta x} = \frac{\mathcal{D}_1(x+\delta x)-\mathcal{D}_1(x)}{\delta x} + O\left(x_1^k Q_{R+2,1}(x)\right).$$

Since

$$\frac{(x+\delta x)Q_{h,v}(x+\delta x) - xQ_{h,v}(x)}{\delta x} = x \cdot \frac{Q_{h,v}(x+\delta x)-Q_{h,v}(x)}{\delta x} + Q_{h,v}(x+\delta x),$$

and

$$\frac{Q_{h,v}(x+\delta x)-Q_{h,v}(x)}{\delta x} = Q'_{h,v}(x) + O\left((\delta x)Q''_{h,v}(x)\right)$$

$$Q'_{h,v}(x) = \frac{1}{x}\, e(\alpha\gamma x_1^{1/\alpha}) \left\{ \gamma x_1^{\frac{1}{\alpha}-1-\frac{\eta_{h,v}}{\beta+1}} - \frac{\eta_{h,v}}{\beta+1}\, x_1^{-\frac{\eta_{h,v}}{\beta+1}-1} \right\} \tag{59}$$

furthermore that

$$Q''_{h,v}(x) = O\left(\frac{1}{x^2}\, e(\alpha\gamma x_1^{\frac{1}{\alpha}-1}\right),$$

with some further simple computation, we obtain the following theorem.

Theorem 2. *Assume that the conditions stated in Theorem 1 are satisfied. Then*

$$\frac{S(x)}{x} = \exp(\alpha\gamma x_1^{1/\alpha}) \left\{ \sum_{\eta_{h,v}\in S_R} H(\eta_{h,v}\Lambda_{h,v} \cdot x_1^{-\frac{\eta_{h,v}}{\beta+1}} + O\left(x^{-\frac{R+2+\frac{\beta}{2}}{\beta+1}}\right) \right\}$$

where

$$\Lambda_{h,v} = 1 + \gamma \cdot x_1^{\frac{1}{\alpha}-1} - \frac{\eta_{h,v}}{\beta_1}\, x_1^{-1}.$$

6.1 Some examples

It is not difficult to give examples of $e(n)$ that satisfy the requirements of the theorems or their extensions mentioned in the Remark. We content ourselves with the following examples

(i) $e(n) = 1$ for all integers $n \geq 1$.

(ii) Let $k \in N$ be fixed. Then there exists a positive integer N_k such that $d_k(n) < n$ for all $n > N_k$. Take

$$e(n) = \begin{cases} d_k(n) & \text{if} \quad n > N_k \\ 1 & \text{if} \quad n \leq N_k. \end{cases}$$

In particular, we can take

$$e(n) = \begin{cases} d(n), & n > 2 \\ 1, & n \leq 2. \end{cases}$$

Note that we have

$$\zeta^k(s) = \frac{1}{(s-1)^k} + \frac{b_1}{(s-1)^{k-1}} + \cdots + \frac{b_{k-1}}{s-1} + u_1(n)$$

and hence we can apply the Theorem 1 in the extended version given in the remark that followed the theorem.

(iii) Take a fixed integer $M > 1$ and let integers $\ell_1, \ell_2, \ldots, \ell_k < M$ satisfy

$$1 \leq \ell_1 < \ell_2 < \cdots < \ell_k < M.$$

Define

$$e(n) = \begin{cases} 1 & \text{if} \quad n \equiv \ell_i (\text{mod } i), \quad i = 1, \ldots, k \\ 0 & \text{otherwise.} \end{cases}$$

Then

$$\log F(s) = \frac{k}{M}\, \zeta(s) + u(s),$$

where $u(s)$ is bounded around $s = 1$.

(iv) Take $e(n)$ to be the completely multiplicative function defined by $e(p) = 1$ if $p \equiv 1 (\bmod 4)$, and $= 0$ if $p \not\equiv 1 \bmod 4$. Let $\chi(n)$ be the Dirichlet character defined by

$$\chi(p) = \begin{cases} 1, & p \equiv 1 (\bmod 4) \\ -1, & p \equiv -1 \bmod 4 \\ 0, & p \equiv 2, \end{cases}$$

Write as usual

$$L(s, \chi) = \sum \frac{\chi(n)}{n^s}.$$

Then

$$\begin{aligned} \zeta(s) L(s, \chi) &= \prod_{p \equiv 1 (\bmod 4)} \left(1 - \frac{1}{p^s}\right)^{-2} \cdot \prod_{p \equiv -1 (\bmod 4)} \left(1 - \frac{1}{p^{2s}}\right)^{-1} \cdot (1 - 2^{-s}) \\ &= U^2(s) u_1(s), \end{aligned}$$

where, as usual, $U(s) = \sum \frac{e(n)}{n^s}$ and $u_1(s)$ is regular if $\operatorname{Re} s > 1/2$. Thus,

$$\begin{aligned} U(s) &:= \left(\frac{\zeta(s) L(s, \chi)}{u_1(s)}\right)^{1/2} \\ &= \frac{1}{(s-1)^{1/2}} \left\{\frac{\zeta(s)(s-1) L(s, \chi)}{u_1(s)}\right\} \\ &= \frac{1}{(s-1)^{1/2}} \cdot u_2(s), \text{ (say)}, \end{aligned}$$

where $u_2(s)$ is regular and $\neq 0$ in $|s - 1| \leq 1/2$. Hence we can write

$$U(s) = \frac{e_0}{(s-1)^{1/2}} + u_3(s),$$

where $u_3(s)$ is bounded in $|s - 1| < 1/2$.

7 Concluding Remarks

In response to a question concerning an asymptotic estimate for $\sum_{n \leq x} q^*(n)$ which the second author raised in a letter to Professor K. Ramachandra, the latter replied in his letter dated March 19, 2004 that his joint paper [2] may be used as follows.

Let $F(s) = \exp \zeta(s) = \exp\left(\sum_{n=1}^{\infty} n^{-s}\right) = e \prod_{n=1}^{\infty} (1 + n^{-s}) \psi(s)$ where $\psi(s)$ is bounded in $\sigma > 1/2$ and analytic there. Put $F(s) = \sum_{n=1}^{\infty} a_n / n^s$. Then using Theorem 6 of [2], we get

$$\sum_{n \leq x} a_n = \frac{1}{2\pi i} \int_{|s-1| = \frac{1}{100}} \frac{F(s) x^s}{s} \, ds + \Omega(x^{1-\varepsilon})$$

for every fixed positive ε.

Of course, we can replace $F(s)$ in the above by

$$\phi(s) = \prod_{n=2}^{\infty}(1 + n^{-s}) = 1 + \sum_{n=2}^{\infty} q^*(n)n^{-s}.$$

We should remark that many other omega results of the above kind can be obtained using the many choices for $F(s)$ given in their paper [2].

In particular, as Professor Ramachandra remarks, in his letter to the second author, we can take in the above result, among many other choices, $F(s)=\exp(\alpha(\zeta(s))^{\beta})$. where α, β are any non-zero complex numbers. Thus one can obtain an omega result for the summatory function generated by $\Pi(1 - d'(n)n^{-s})$, where $d'(n)$, for $n > 2$, denotes the number of divisors of $n > 2$, with $d'(2) = 1$.

The richness and variety of possible applications of the results of [2] in obtaining omega results is further illustrated by Schinzel's paper [9]. Using the results of some of the lemmas of [1] — which contains an earlier version of [2] — Schinzel obtained an Ω-estimate for the error term in the summatory function for $r^2(n)$, where $r(n)$ denotes the number of representations of n as a sum of two square integers. It should be noted that an asymptotic formula for $\sum_{n \leq x} r^2(n)$, with an O-term estimate for the error term has been obtained by Sierpinski as far back as 1908. This O-term has since been improved by several mathematicians, including S. Ramanujan (see [9] for details).

But, our purpose in this paper is not to get Ω-results for the summatory product partition functions, but asymptotic formulas with O-estimates for the error term.

Final Remark

Using the approach of this paper, we can extend the asymptotic result (1.1) of Oppenheim [8] and Szekeres - Turan [11] and obtain more general results by studying the Dirichlet series expansion of $\prod_{n=2}^{\infty}(1 - \frac{g(n)}{n^s})^{-1}$ for certain types of $g(n) \geq 0$. We shall not go into details here.

Acknowledgement

The second author is extremely thankful to Ms. Vivian Spak for technical help without which the preparation of this paper would not have been possible.

References

[1] R. Balasubramanian, K. Ramachandra and M.V. Subbarao, On the error function in the asymptotic formula for the counting function of k-full numbers, *Acta Arith.* **50** (1988), 107–108.

[2] R. Balasubramanian and K. Ramachandra, Some problems in analytic number theory - IV, *Hardy-Ramunujan J.*, **25** (2002), 5–21.

[3] R. Canfied, Paul Erdös and Carl Pomerance, On the problem of Oppenheim concerning "Factorisatio Numerorum", *J. Number Theory*, **17** (1983), 1–28.

[4] N.G. de Bruijn, Asymptotic Methods in Analysis, corrected reprint of the third edition, Dover Publications, Inc., New York, 1981.

[5] Ryuji Kanejwa, On the multiplicative partition function, *Tsukuha J. Math.* **7** (1983), no. 2, 355–365.

[6] P.A. MacMahon, Dirichlet series and the theory of partitions, *Proc. London Math. Soc.* (2) **22** (1924), 404–411;

[7] A. Oppenheim, On an arithmetic function, *Proc. London Math. Soc.* **I** (1926), 205–211.

[8] A. Oppenheim, On an arithmetic function, II, *J. London Math. Soc.* **II** (1927), 123–130.

[9] A. Schinzel, On an analytic problem considered by Sierpinski and Ramanujan, *New Trends in Probability and Statistics*, F. Schweiger and E. Manstarius (eds.), 1992, VSP/TEV, 165–171.

[10] M.V. Subbarao and A. Verma, Some remarks on a product expansion: an unexplored partition function, Proc. Conf. Symbolic Computation, Number Theory, Special Functions, Physics and Combinatorics, held in Gainesville, Florida 1999, F.G. Garvan and M.E.H. Ismail (eds.), Kluwer Academic Publishers, 2001, pp. 267–283.

[11] G. Szekeres and P. Turan, Über das Zweite Hauptproblem der "Factorisatio Numerorium", *Acta Litt. Sci. Szeged* **6** (1933), 143–154.

I. Kátai
Department of Computer Angebra,
Eotvös Loránd University,
Pázmány PÃter SÃt 1/C,
H-1117 Budapest, Hungary,
e-mail: katai@compalg.inf.elte.hu

M.V. Subbarao
Department of Mathematical and Statistical Sciences,
University of Alberta,
Edmonton, Alberta, Canada, T6G 2G1
e-mail: m.v.subbarao@ualberta.ca

The Riemann Zeta Function and Related Themes – 2006, pp. 115–120

The distribution of integers with given number of prime factors in almost all short intervals

I. Kátai[*†] and M.V. Subbarao[†‡]

Dedicated to Professor K. Ramachandra on his 70th birthday

§1. By using the method of Ramachandra [1], in [2] and [3] the asymptotic of the sums

$$\sum_{n\in[x,x+h]} z^{\omega(n)}, \qquad \sum_{n\in[x,x+h]} z^{\omega(n)}|\mu(n)|$$

has been found for $h \geq x^{7/12+\varepsilon}$, where $z \in \mathbb{C}$, $|z| \leq B$. Here B is an arbitrary constant.

Applying the second assertion of Ramachandra [1], one can get that

$$\frac{1}{X}\int_X^{2X} \left|E(x+h) - E(x) - \frac{h}{X}E(X)\right|^2 dx \ll$$

$$\ll h^2 e^{-(\log x)^{1/6}} + X^{2\varphi'}, \tag{1}$$

where $\varphi' = 1/6+\varepsilon$ ($\varepsilon > 0$ is an arbitrary positive constant), and $E(x)$ is one of the following functions

$$E_1(x) = \sum_{n\leq x} z^{\omega(n)}, \quad |z| \leq B, \tag{2}$$

$$E_2(x) = \sum_{n\leq x} z^{\omega(n)}|\mu(m)|, \quad |z| \leq B, \tag{3}$$

$$E_3(x) = \sum_{n\leq x} z^{\Omega(n)}, \quad |z| \leq 2-\delta, \quad \delta > 0, \text{ constant.} \tag{4}$$

By using the theorem of A. Selberg [4] (see also Lemma 9.2 in Kubilius [5]) we obtain that

$$\frac{E_1(x)}{x} = \varphi(z)(\log x)^{z-1} + O\left((\log x)^{z-2}\right),$$
$$\varphi(z) = \frac{1}{\Gamma(z)}\prod_p \left(1-\frac{1}{p}\right)^z \left(1+\frac{z}{p-1}\right), \tag{5}$$

$$\frac{E_2(x)}{x} = \psi(z)(\log x)^{z-1} + O\left((\log x)^{z-2}\right),$$
$$\psi(z) = \frac{1}{\Gamma(z)}\prod_p \left(1-\frac{1}{p}\right)^z \left(1+\frac{z}{p}\right), \tag{6}$$

---

[*]Financially supported by an NSERC grant of the second named author.
[†]Partially supported by OTKA T 46993 and the fund of Applied Number Theory Group of HAS
[‡]Editors regret to record the passing away of Prof. M. V. Subbarao on February 15, 2006.

$$\frac{E_3(x)}{x} = G(z)(\log x)^{z-1} + O\left((\log x)^{z-2}\right),$$
$$G(z) = \frac{1}{\Gamma(1+z)} \prod_p \left(1 - \frac{z}{p}\right)^{-1} \left(1 - \frac{1}{p}\right)^z. \tag{7}$$

Let

$$\pi_l(A) = \#\{n \,|\in A,\ \omega(n) = l\},$$
$$\Pi_l(A) = \#\{n \mid n \in A,\ \omega(n) = l,\ \mu(n) \neq 0\},$$
$$T_l(A) = \#\{n \mid n \in A,\ \Omega(n) = l\}.$$

Let

$$\xi_r(x) := \frac{1}{(\log x)} \cdot \frac{(\log\log x)^{r-1}}{(r-1)!} \quad (r \in \mathbb{N}).$$

Hence, by using standard techniques, we shall get the following assertions which we formulate now as

Theorem 1. *Let $X^{\frac{1}{6}+\varepsilon} \leq h \leq X$. Then, for $1 \leq l \leq B \log\log X$,*

$$\frac{1}{X} \int_X^{2X} (\pi_l([x, x+h]) - h\xi_l(x))^2 \, dx \ll \frac{\xi_l^2(X)}{\log\log X}, \tag{8}$$

$$\frac{1}{X} \int_X^{2X} \left(\Pi_l([x, x+h]) - \frac{6}{\pi^2}\xi_l(x)\right)^2 dx \ll \frac{\xi_l^2(X)}{\log\log X}, \tag{9}$$

and for $1 \leq l \leq (2-\delta)\log\log X$

$$\frac{1}{X} \int_X^{2X} \left(T_l([x, x+h]) - \frac{hT_l(X)}{X}\right)^2 \ll \left(\frac{T_l(X)}{X}\right)^2 \frac{1}{\log\log X} \tag{10}$$

§2. Let $q \geq 2$ be a fixed integer. The q-ary expansion of n is denoted by

$$n = \sum_{\nu=0}^{\infty} \varepsilon_\nu(n) q^\nu,$$

where $\varepsilon_\nu(n) \in \{0, 1, \ldots, q-1\} = A_q$. A function $g : \mathbb{N}_0 \to \mathbb{C}$ is said to be q-multiplicative, if $g(0) = 1$, and

$$g(n) = \prod_{j=0}^{\infty} g(\varepsilon_j(n) q^j)$$

holds for every $n \in \mathbb{N}$. Let $\alpha_q(n) = \sum \varepsilon_j(n)$, $\alpha_q(n)$ is the so called sum of digits function.

$$\frac{1}{\pi(x)} \#\{p \leq x \mid \alpha_2(p) \equiv 0 (\text{mod } 2)\} \to 1/2 \quad (x \to \infty)$$

where p runs over the set of primes. Presently it is not known whether there exist such infinite sequences of primes $\{p_j\}, \{q_j\}$ such that $\alpha_2(p_j)$ =even, $\alpha_2(q_j)$ =odd for $j = 1, 2, \ldots$.

Quite plausible is to guess that if g is q-multiplicative, $|g(n)| == 1 \quad (n \in \mathbb{N})$, $g(p) = 1$ for every large prime p, then $g(n) = 1$ identically.

Recently, in a paper [6] the following assertion has been proved:

There exists a constant c_1 such that, if g is q-multiplicative, $g(p) =$ constant for every large prime p, then $g^k(nq) = 1$ holds for every $n \in \mathbb{N}$, and for a suitable fixed k, in $1 \le k \le c_1$.

We can prove the following

Theorem 2. *There exists a constant c_2 with the following property. Let g be a q-multiplicative function, $|g(n)| = 1$, and B be a constant. Assume that there exists a sequence of integers $N_1 < N_2 < \ldots$ and a sequence of integers $k_\nu \in [1, B \log N_\nu]$, and α_ν such that*

$$\begin{cases} g(\pi) = \alpha_\nu & \text{if } \pi \in [q^{N_\nu}, 4q^{N_\nu}], \\ & \pi \in \mathcal{P}_{k_\nu}, \end{cases} \tag{11}$$

where $\mathcal{P}_k$ is the set of square-free numbers with exactly k prime factors.

Then there exists some integer $r \in [1, c_2]$ such that $g^r(nq) = 1 \quad (n \in \mathbb{N})$.

Proof. The proof is very similar to that of the theorem in [6].

If $k_\nu = 1$ holds for infinitely many ν, then by a small modification of the method used in [6], we obtain the theorem.

Assume that $k_\nu \ge 2$ for $\nu \ge \nu_0$. Let $\nu \ge \nu_0$ be fixed, and put

$$k_\nu = k\ (\ge 2),\ N_\nu = N,\ S := \left[\frac{N}{10}\right],\ R = \left[\frac{N}{3}\right].$$

Let

$$\begin{aligned} \mathcal{B} \ &= \ \Big\{\pi_{k-1}p,\ (\pi_{k-1}, q) = 1,\ \pi_{k-1} \in \left[q^S, 2q^S\right], \\ & \qquad p \in \left[q^{N-S}, 2q^{N-S}\right],\ p \in \mathcal{P},\ \pi_{k-1} \in \mathcal{P}_{k-1}\Big\}. \end{aligned}$$

For some interval I let $B(I)$ be the number of elements of $\mathcal{B} \cap I$. It is clear that

$$\#(\mathcal{B}) = q^S \cdot \xi_{k-1}(q^S) \cdot \frac{q^{N-S}}{\log q^{N-S}} \left(1 + O\left(\frac{1}{N}\right)\right),$$

since

$$\pi_{k-1}\left(\left[q^S, 2q^S\right]\right) = q^S \cdot \xi_{k-1}(q^S)\left(1 + O\left(\frac{1}{N}\right)\right).$$

Let $u_0 = q^{N-R}$, $u_1 = 2 \cdot q^{N-R}$, $J_u := \left[u \cdot q^R,\ (u+1)q^R\right)$.

Then, for $u \in [u_0, u_1 - 1]$ we have

$$B(J_u) = \sum_{q^S \le \pi_{k-1} \le 2q^S} \pi\left(\left[\frac{u \cdot q^R}{\pi_{k-1}}, \frac{(u+1) \cdot q^R}{\pi_{k-1}}\right)\right)$$

where $\pi(J) =$ number of primes located in the interval J.

Thus

$$\left| B(J_u) - \frac{q^R}{\log q^{N-S}} \sum_{\pi_{k-1} \in [q^S, 2q^S)} \frac{1}{\pi_{k-1}} \right| \leq$$

$$\leq \sum_{\pi_{k-1} \in [q^S, 2q^S)} \left| \pi \left(\left[\frac{u \cdot q^R}{\pi_{k-1}}, \frac{(u+1)q^R}{\pi_{k-1}} \right) \right) - \frac{q^R}{\pi_{k-1} \log q^{N-S}} \right|. \tag{12}$$

Let

$$\psi_{k,S} = \sum_{\pi_{k-1} \in [q^S, 2q^S)} \frac{1}{\pi_{k-1}}.$$

It is clear that

$$\psi_{k,S} = (\log 2)\xi_{k-1}\left(q^S\right)\left(1 + O\left(\frac{1}{N}\right)\right) + (\log 2)\xi_{k-2}\left(q^S\right)\left(1 + O\left(\frac{1}{N}\right)\right).$$

From (12) we obtain that

$$\left| B(J_u) - \frac{q^R}{\log q^{N-S}} \cdot \psi_{k,S} \right|^2 \leq$$

$$\leq \pi_{k-1}(2q^S) \sum_{q^S \leq \pi_{k-1} \leq 2q^S} \left| \pi \left(\left[\frac{u \cdot q^R}{\pi_{k-1}}, \frac{(u+1)q^R}{\pi_{k-1}} \right) \right) - \frac{q^R}{\pi_{k-1} \log q^{N-S}} \right|^2. \tag{13}$$

By using the theorem of A. Selberg [4], according to

$$\frac{1}{X} \int_X^{2X} |\psi(x+h) - \psi(x) - h|^2 \, dx = o(1), \tag{14}$$

if $X > j \geq X^{\frac{19}{77}+\varepsilon}$. From the theorem of Ramachandra (13) follows for $X \geq h \geq x^{1/6+\varepsilon}$.

Summing (13) over $u = u_0, \ldots, u_1$, and taking into account (14), we deduce

$$E := \sum_{u=u_0}^{u_1} \left| B(J_u) - \frac{q^R}{\log q^{N-s}} \cdot \psi_{k,s} \right|^2 \leq \pi_{k-1}(2q^s) \sum_{\pi_{k-1} \in [q^s, 2q^s]} A(\pi_{k-1}),$$

where

$$A(\pi_{k-1}) = \sum_{u=u_0}^{u_1} \left| \pi \left(\left[\frac{u \cdot q^R}{\pi_{k-1}}, \frac{(u+1)q^R}{\pi_{k-1}} \right) \right) - \frac{q^R}{\pi_{k-1} \log q^{N-S}} \right|^2.$$

From (14), we obtain that

$$A(\pi_{k-1}) \ll \varepsilon_N \cdot \left(\frac{q^R}{\pi_{k-1} \log q^{N-S}} \right)^2 \cdot q^{N-R},$$

where $\varepsilon_N \to 0 \quad (n \to \infty)$, consequently

$$E \ll \varepsilon_N \cdot q^S \xi_{k-1}(q^S) q^{2R+N-R} \cdot \frac{1}{N^2} \sum \frac{1}{\pi_{k-1}^2} \ll \frac{\varepsilon_N}{N^2} \cdot q^{N+R} \cdot \xi_{k-1}^2(q^S).$$

If $b_1, b_2 \in J_u \cap \mathcal{B}$, $b_j \equiv l_j \pmod{q^R}$, then $1 = g(b_1)\overline{g}(b_2) = g(l_1)\overline{g}(l_2)$. Consequently, $g(l_1) = g(l_2)$, if there exists $b_1, b_2 \in \mathcal{B}$, such that $b_1 \equiv l_1 \pmod{q^R}$, $b_2 - b_1 = l_2 - l_1$.

Let

$$\kappa(q^N, l_1, l_2) = \#\{b_1, b_2 \in \mathcal{B},\ b_1 \equiv l_1 \pmod{q^R},\ b_2 - b_1 = l_2 - l_1\}.$$

Let $g(b) = \alpha$ for every $b \in \mathcal{B}$. Then

$$\begin{aligned} \sum\nolimits_0 &= \sum g(b) = \alpha\#(\mathcal{B}), \\ \#(\mathcal{B}) &= \left(\pi\left(2q^{N-S}\right) - \pi\left(q^{N-S}\right)\right)\pi_{k-1}\left(\left[q^S, 2q^S\right]\right) \\ \sum\nolimits_0 &= \sum_{u=u_0}^{u_1}\sum_{b\in J_u} g(b), \end{aligned}$$

$$|\#(\mathcal{B})|^2 \le \left\{\sum_u 1\right\}\left\{\sum_{u=u_0}^{u_1}\left|\sum_{b\in J_u} g(b)\right|^2\right\}.$$

The second sum on the right hand side is not larger than

$$2\sum_{\substack{l_1<l_2<q^R \\ g(l_1)=g(l_2)}} \kappa(q^N, l_1, l_2) + \#(\mathcal{B}).$$

Thus

$$\#(\mathcal{B})^2 \le q^{N-R}\#(\mathcal{B}) + 2q^{N-R}\sum_{\substack{l_1<l_2<q^R \\ g(l_1)=g(l_2)}} \kappa(q^N, l_1, l_2). \tag{15}$$

Let us give an upper estimation for the last sum on the right hand side of (15). Let $l_1 < l_2 < q^R$, $(l_1 l_2, q) = 1$

$$\tau(l_1, l_2) := \#\{b_2 - b_1 = l_2 - l_1,\ b_1 \equiv l_1 \pmod{q^R}\quad b_1, b_2 \in \mathcal{B}\}.$$

For fixed $\pi_{k-1}^{(1)}, \pi_{k-1}^{(2)}$, the number of prime pairs p_1, p_2 for which

$$\begin{aligned} \pi_{k-1}^{(2)}p_2 - \pi_{k-1}^{(1)}p_1 &= l_2 - l_1, & p_1\pi_{k-1}^1 &\in \mathcal{B} \\ p_1 &\equiv l_1(q^R), & p_2\pi_{k-1}^2 &\in \mathcal{B} \end{aligned} \tag{16}$$

is less than

$$\frac{cq^{N-2s}}{q^R N^2},$$

with a suitable numerical constant c. Since the number of possible choices $\pi_{k-1}^{(1)}$, $\pi_{k-1}^{(2)}$ is $\le \#^2\left\{\pi_{k-1} \in \left[q^S, q^{2S}\right]\right\} \le cq^{2S}\xi_{k-1}^2(q^S)$, thus

$$\tau(l_1, l_2) \le c\frac{q^N}{q^R N^2}\xi_{k-1}^2(q^S),$$

whence we obtain that

$$\#\{l_1 < l_2 < q^R \mid g(l_1) = g(l_2)\} > c^* q^{2R}.$$

Hence we can deduce our theorem on the same way, as in [6].

References

[1] K. Ramachandra, Some problems on analytic number theory, *Acta Arithmetica* **31** (1976), 313–324.

[2] I. Kátai, A remark on a paper of Ramachandra, in: Number Theory, *Proc. Ootacamund*, K. Alladi (Ed.), Lecture Notes in Math. 1122 Springer (1984), pp. 147–152.

[3] I. Kátai and M.V. Subbarao, Some remarks on a paper of Ramachandra, *Liet. Matem. Rink.* **43**(4), (2003), 497–506.

[4] A. Selberg, A note on a paper by L.G. Sathe, *J. Indian Math. Soc.* **18**, (1954), 83–87.

[5] J.P. Kubilius, *Probabilistic methods in the theory of numbers* (in Russian), Vilnius (1959).

[6] I. Kátai and K.H. Indlekofer, On q-multiplicative functions taking a fixed value on the set of primes, *Periodica Math. Hung.* **42** (2001), 45–50.

I. Kátai
Department of Computer Angebra,
Eotvös Loránd University,
Pázmány PÃter SÃt 1/C,
H-1117 Budapest, Hungary,

e-mail: katai@compalg.inf.elte.hu

M.V. Subbarao
Department of Mathematical and Statistical Sciences,
University of Alberta,
Edmonton, Alberta, Canada, T6G 2G1

e-mail: m.v.subbarao@ualberta.ca

The Riemann Zeta Function and Related Themes – 2006, pp. 121–129

On nonvanishing infinite sums

N. Saradha

Dedicated to Professor K. Ramachandra on his 70th birthday

1 Introduction

Let $1 < q = p_1^{\alpha_1} \dots p_r^{\alpha_r}$ with $p_1 < \dots < p_r, p_i$ prime and $\alpha_i \geq 1$ for $1 \leq i \leq r$. We denote by $\varphi(q)$ and $\omega(q)$, the Euler totient function of q and the number of distinct prime divisors of q, respectively. Suppose f is an arithmetic function, periodic mod q defined by

$$f(n) = \begin{cases} \pm 1 \text{ if } n \text{ is not a multiple of } q \\ 0 \text{ otherwise.} \end{cases}$$

Then Erdős (see [2]), posed the problem whether the infinite series

$$S = \sum_{n=1}^{\infty} \frac{f(n)}{n} \tag{1}$$

can vanish whenever convergent ? A necessary and sufficient condition for the series to converge is that

$$\sum_{n=1}^{q} f(n) = \sum_{n=1}^{q-1} f(n) = 0.$$

Hence if q is even, the series S does not converge. Thus we restrict to q *odd.* In [5], Okada showed that the series S does not vanish provided

$$2\varphi(q) \geq q. \tag{2}$$

This condition is satisfied for all q with $\omega(q) \leq 2$. In particular, when q is a prime then S does not vanish. By a result of Adhikari, Saradha, Shorey and Tijdeman [1], the non-vanishing of S implies that S is transcendental. Their result was a consequence of Baker's theory on linear forms in logarithms and a theorem of Baker, Birch and Wirsing [4] in which they derived a necessary and sufficient condition for the nonvanishing of the series

$$T = \sum_{n=1}^{\infty} \frac{g(n)}{n}$$

where g is an algebraic valued, periodic function with period q. Okada [5] translated the necessary and sufficient conditions into a criterion consisting of a set of $\varphi(q)+\omega(q)$ homogeneous linear equations to be satisfied by the values of g. Using Okada's criterion Tijdeman [7] showed that T does not vanish if g is completely multiplicative or if g is multiplicative such that $|g(p^k)| < p - 1$ for every prime divisor p of q and every positive integer k. The criterion of Okada was reformulated by Saradha and Tijdeman ([6], see Lemma 1). Using this, they gave explicit conditions under which some special infinite sums do not vanish. In this paper we shall use the reformulated criterion of [6] to study the nonvanishing of S when $\omega(q) = 3, 4$. We show

Theorem 1. (i) *Let* $\omega(q) = 3$. *Then* S *does not vanish except possibly when*

$$(p_1, p_2, p_3) \in \{(3,5,7), (3,5,11), (3,5,13)\}$$

and $q \notin A$ *where* A *is given by*

$$A = \{3\cdot5\cdot7, 3^2\cdot5\cdot7, 3\cdot5\cdot11, 3\cdot5\cdot13, 3\cdot5^2\cdot11, 3\cdot5^2\cdot13, 3^2\cdot5\cdot11, \quad (3)$$
$$3^2\cdot5\cdot13, 3^2\cdot5^2\cdot11, 3^2\cdot5^2\cdot13, 3^3\cdot5\cdot11, 3^3\cdot5\cdot13, 3^3\cdot5^2\cdot11, 3^3\cdot5^2\cdot13\}.$$

(ii) *Let* $\omega(q) = 4$. *Then* S *does not vanish except possibly when*

$$(p_1, p_2, p_3, p_4) = (3,5,7,p_4); (3,5,11,p_4); (3,5,13,p_4); (3,5,17,p_4) \text{ with } p_4 \le 251;$$
$$(3,5,19,p_4) \text{ with } p_4 \le 89; (3,5,23,p_4) \text{ with } p_4 \le 47; (3,5,29,p_4) \text{ with } p_4 \le 31;$$
$$(3,7,11,p_4) \text{ with } p_4 \le 23; (3,7,13,p_4) \text{ with } p_4 \le 19$$

and q *not listed in Table* 1.

As a consequence of Theorem 1, we get

Corollary. *Let* $q \notin \{525, 735, 945\}$. *Then* S *does not vanish for all* $q \le 1154$.

We derive Theorem 1 from the following result.

Theorem 2. *Let* $h = \max_{1\le i\le r}(p_i^{\alpha_i})$. *Assume that*

$$2\varphi(q) \ge q\left(1 - \frac{1}{h}\right). \quad (4)$$

Then S *does not vanish.*

The values of q in (3) and Table 1 for which S does not vanish are not covered by the result of Okada. The nonvanishing of S when $q = 525, 735, 945$ is not known. For a survey on related problems, refer to [2], [3] and [7].

I wish to thank Professor T.N. Shorey for his suggestions on an earlier draft.

2 Lemmas

We begin with the reformulated criterion of [6, Lemma 1].

Lemma 1. *Let* g *be an algebraic valued, periodic function with period* q. *Let* M *be the set of integers composed only of* $p_1, \dots, p_r$. *Suppose for any integer* $n > 0$, $\nu_p(n)$ *denotes the order of the prime* p *in* n. *Define*

$$\epsilon(r,p) = \begin{cases} \nu_p(r) \text{ if } \nu_p(r) < \nu_p(q) \\ \nu_p(q) + \frac{1}{p-1} \text{ if } \nu_p(r) \ge \nu_p(q) \end{cases}$$

Then

$$T = \sum_{n=1}^{\infty} \frac{g(n)}{n}$$

vanishes if and only if the following two conditions hold.

(i) $\sum_{n \in M} \frac{g(an)}{n} = 0$ *for every* a *with* $1 \leq a < q$ *and* $\gcd(a, q) = 1$

(ii) $\sum^{q}_{\substack{t=1 \\ \gcd(t,q)>1}} g(t)\epsilon(t, p_i) = 0$ *for* $1 \leq i \leq r$.

Lemma 2. *Suppose* S *vanishes. Let* a, b *be two positive integers with* $\gcd(a, q) = \gcd(b, q) = 1$ *and* $a \equiv b \left(\bmod \frac{q}{p_i^{\alpha_i}}\right)$ *for some* i *with* $1 \leq i \leq r$. *Suppose*

$$\frac{q}{\varphi(q)} \frac{p_i^{\alpha_i} - 1}{p_i^{\alpha_i}} < 2. \tag{5}$$

Then $f(a) = f(b)$.

Proof: We apply Lemma 1 with $T = S$. Then (i) and (ii) are valid with $g = f$. From (i), we have

$$f(a') = -\sum_{\substack{m \in M \\ m>1}} \frac{f(a'm)}{m} \text{ for any } a' \text{ with } \gcd(a', q) = 1. \tag{6}$$

Since $a \equiv b \left(\bmod \frac{q}{p_i^{\alpha_i}}\right)$, we have $f(am) = f(bm)$ whenever $p_i^{\alpha_i} \mid m$. Thus it follows from (6) that

$$\begin{aligned} f(a) - f(b) &= \sum_{\substack{m \in M \\ m>1}} \frac{f(bm) - f(am)}{m} \\ &= \sum_{\substack{m \in M, m>1 \\ p_i \nmid m}} \frac{f(bm) - f(am)}{m} + \sum_{\substack{m \in M, m>1 \\ 0<v_{p_i}(m)<\alpha_i}} \frac{f(bm) - f(am)}{m}. \end{aligned}$$

Hence

$$| f(a) - f(b) | \leq 2 \left(\sum_{\substack{m \in M, m>1 \\ p_i \nmid m}} \frac{1}{m} + \sum_{\substack{m \in M, m>1 \\ 0<v_{p_i}(m)<\alpha_i}} \frac{1}{m} \right).$$

Suppose

$$P_i = \left(1 - \frac{1}{p_1}\right)^{-1} \cdots \left(1 - \frac{1}{p_r}\right)^{-1} \left(1 - \frac{1}{p_i}\right) = \frac{q}{\varphi(q)} \frac{p_i - 1}{p_i}.$$

Since M is the set of integers composed only of $p_1, \ldots, p_r$, we see that

$$\sum_{\substack{m \in M, m>1 \\ p_i \nmid m}} \frac{1}{m} = \frac{(1 + \frac{1}{p_1} + \cdots) \ldots (1 + \frac{1}{p_r} + \cdots)}{(1 + \frac{1}{p_i} + \cdots)} - 1 = P_i - 1$$

and

$$\sum_{\substack{m\in M, m>1 \\ 0<\nu_{p_i}(m)<\alpha_i}} \frac{1}{m} = \frac{(1+\frac{1}{p_1}+\cdots)\dots(1+\frac{1}{p_r}+\cdots)}{(1+\frac{1}{p_i}+\cdots)}\left(\frac{1}{p_i}+\cdots+\frac{1}{p_i^{\alpha_i-1}}\right)$$

$$= \frac{P_i}{p_i-1}\left(1-\frac{1}{p_i^{\alpha_i-1}}\right).$$

Thus we find that

$$|f(a)-f(b)| \le 2\left(P_i - 1 + P_i\frac{1-\frac{1}{p_i^{\alpha_i-1}}}{p_i-1}\right) \le 2\left(\frac{q}{\varphi(q)}\frac{p_i^{\alpha_i}-1}{p_i^{\alpha_i}}-1\right) < 2$$

by our supposition. Thus $f(a) = f(b)$. □

Lemma 3. *Assume that S vanishes. Suppose* (5) *holds for every i with* $1 \le i \le r$. *Then*

$$f(a) = f(b)$$

for integers a, b with $\gcd(a,q) = \gcd(b,q) = 1$.

Proof: It is enough to prove the result for a, b with $1 \le a, b < q$ and $\gcd(a,q) = \gcd(b,q) = 1$. There are $\varphi(q)$ number of such a's. For every i with $1 \le i \le r$, we partition the a's into $\varphi(\frac{q}{p_i^{\alpha_i}})$ classes, say

$$C_1^{(i)}, \dots, C_{\varphi(q/p_i^{\alpha_i})}^{(i)},$$

each class consisting of $\varphi(p_i^{\alpha_i})$ elements such that if $a, b \in C_h^{(i)}$ for some h with $1 \le h \le \varphi(q/p_i^{\alpha_i})$, then $a \equiv b(\text{mod } q/p_i^{\alpha_i})$. Hence by Lemma 2, we have

$$f(a) = f(b) \text{ if } a, b \in C_h^{(i)} \text{ with } 1 \le h \le \varphi(q/p_i^{\alpha_i}). \tag{7}$$

We set

$$X_i = \left\{C_1^{(i)}, \dots, C_{\varphi(q/p_i^{\alpha_i})}^{(i)}\right\} \quad \text{for} \quad 1 \le i \le r.$$

Suppose $a, b \in C_{h_1}^{(i)}$ for some $0 \le h_1 \le \varphi(q/p_i^{\alpha_i})$ and $a \in C_{h_2}^{(j)}$ for some $0 \le h_2 \le \varphi(q/p_j^{\alpha_j})$ with $0 \le i < j \le r$. Then we claim that $b \notin C_{h_2}^{(j)}$. For, otherwise, we have $a \equiv b(\text{mod } q/p_j^{\alpha_j})$. Also $a \equiv b(\text{mod } p_j^{\alpha_j})$ since $a, b \in C_{h_1}^{(i)}$. Hence $a \equiv b(\text{mod } q)$, a contradiction which proves the claim. Hence $\varphi(p_i^{\alpha_i})$ elements in $C_{h_1}^{(i)}$ fall into different classes of X_j. (Note that $\varphi(p_i^{\alpha_i}) \le \varphi(q/p_i^{\alpha_j})$). Thus there are

$$\varphi(p_i^{\alpha_i})\varphi(p_j^{\alpha_j})$$

elements coprime to q for which f takes the same value. Starting from $p_1^{\alpha_1}$ and moving up to $p_r^{\alpha_r}$, we find that there are

$$\varphi(p_1^{\alpha_1})\dots\varphi(p_r^{\alpha_r}) = \varphi(q)$$

elements, coprime to q for which f takes the same value. □

Lemma 4. *Suppose that S vanishes. Assume that* (5) *holds for every i with $1 \le i \le r$. Then condition* (ii) *of Lemma* 1 *does not hold for any $p = p_i$.*

Proof: Suppose S vanishes and (5) holds for every i with $1 \le i \le r$. Then Lemmas 1 and 3 are valid. Since $\sum_{h=1}^{q} f(h) = 0$, we find that

$$\sum_{\substack{a=1\\ \gcd(a,q)=1}}^{q} f(a) = -\sum_{\substack{h=1\\ \gcd(h,q)>1}}^{q} f(h). \tag{8}$$

By Lemma 3, we get

$$\left|\sum_{\substack{a=1\\ \gcd(a,q)=1}}^{q} f(a)\right| = \varphi(q). \tag{9}$$

We write the set of integers $\le q$ and divisible by p_i as

$$S^{(i)} = \cup_{j=1}^{\alpha_i} S_j^{(i)}$$

where $S_j^{(i)}$ for $1 \le j < \alpha_i$ denotes those integers $\le q$ which are exactly divisible by p_i^j. Further, $S_{\alpha_i}^{(i)}$ denotes all integers $\le q$ which are divisible by $p_i^{\alpha_i}$. Then condition (ii) of Lemma 1 gives

$$\sum_{h\in S_1^{(i)}} f(h) + 2\sum_{h\in S_2^{(i)}} f(h) + \cdots + (\alpha_i - 1)\sum_{h\in S_{\alpha_i-1}^{(i)}} f(h) + \left(\alpha_i + \frac{1}{p_i - 1}\right)\sum_{h\in S_{\alpha_i}^{(i)}} f(h)$$
$$= 0 \text{ for } 1 \le i \le r. \tag{10}$$

If further, p_i exactly divides q, then

$$\sum_{h\in S^{(i)}} f(h) = 0. \tag{11}$$

Now we consider the right hand side of (8) and apply (10) and (11) to get

$$\sum_{\substack{h=1\\ \gcd(h,q)>1}}^{q} f(h) = -\sum_{\substack{h=1\\ \gcd(h,q)>1, h\notin S^{(i)}}}^{q} f(h) - \sum_{\substack{h=1\\ \gcd(h,q)>1, h\in S^{(i)}}}^{q} f(h) = -\sum_{\substack{h=1\\ \gcd(h,q)>1, h\notin S^{(i)}}}^{q} f(h)$$
$$+\left(\sum_{h\in S_2^{(i)}} f(h) + \cdots + (\alpha_i - 2)\sum_{h\in S_{\alpha_i-1}^{(i)}} f(h) + \left(\alpha_i - 1 + \frac{1}{p_i - 1}\right)\sum_{h\in S_{\alpha_i}^{(i)}} f(h)\right)\delta_i$$

where

$$\delta_i = \begin{cases} 0 \text{ if } p_i \parallel q \\ 1 \text{ otherwise.} \end{cases}$$

Thus

$$\left|-\sum_{\substack{h=1\\ \gcd(h,q)>1}}^{q} f(h)\right| \le q-\varphi(q)-\frac{q}{p_i}+\delta_i\left(\frac{q}{p_i^2}-\frac{q}{p_i^3}+2\left(\frac{q}{p_i^3}-\frac{q}{p_i^4}\right)+\cdots\right.$$
$$\left.+(\alpha_i-2)\left(\frac{q}{p_i^{\alpha_i-1}}-\frac{q}{p_i^{\alpha_i}}\right)+\left(\alpha_i-1+\frac{1}{p_i-1}\right)\frac{q}{p_i^{\alpha_i}}\right)$$
$$=q-\varphi(q)-\frac{q}{p_i}+\delta_i\left(\frac{q}{p_i^2}+\frac{q}{p_i^3}+\cdots+\frac{q}{p_i^{\alpha_i}}+\frac{1}{p_i-1}\frac{q}{p_i^{\alpha_i}}\right)$$
$$=q-\varphi(q)-\frac{q}{p_i}+\delta_i\left(\frac{q(p_i^{\alpha_i-1}-1)}{p_i^{\alpha_i}(p_i-1)}+\frac{q}{p_i^{\alpha_i}(p_i-1)}\right)$$
$$=q-\varphi(q)-\frac{q}{p_i}+\delta_i\frac{q}{p_i(p_i-1)}$$

which by (8) and (9) gives

$$2\varphi(q)\le q\left(1-\frac{1}{p_i}+\frac{\delta_i}{p_i(p_i-1)}\right). \tag{12}$$

When $\delta_i=0$, by (5) and (12), we have

$$2\le\frac{q}{\varphi(q)}\left(1-\frac{1}{p_i}\right)<2,$$

a contradiction. Let $\delta_i=1$. Then $\alpha_i>1$ and

$$2\le\frac{2p_i^{\alpha_i}}{(p_i^{\alpha_i}-1)}\left(1-\frac{1}{p_i}+\frac{1}{p_i(p_i-1)}\right)$$

implying

$$p_i^{\alpha_i-1}-1\le\frac{p_i^{\alpha_i-1}}{p_i-1}$$

which is impossible. □

Suppose we have

$$\alpha\le\frac{\varphi(q)}{q}\le\beta. \tag{13}$$

We define

$$s_0=s_0'=1,\, s_i=\prod_{j=1}^{i}\left(1-\frac{1}{p_j}\right)\text{ for } i\ge 1.$$

Since

$$\frac{\varphi(q)}{q}=\left(1-\frac{1}{p_1}\right)\cdots\left(1-\frac{1}{p_i}\right)\cdots\left(1-\frac{1}{p_r}\right), \tag{14}$$

we see that

$$\left(1-\frac{1}{p_i}\right)^{r-i+1}s_{i-1}\le\beta.$$

Thus we get

$$p_i < \frac{1}{1-(\frac{\beta}{s_{i-1}})^{1/(r-i+1)}} \text{ for } 1 \le i \le r. \tag{15}$$

The following lemma is a consequence of (13) with $\beta = 1/2$.

Lemma 5. *Suppose $2\varphi(q) \le q$. Then*

(i) *If $\omega(q) = 3$, then $(p_1, p_2, p_3) = (3,5,7)$ or $(3,5,11)$ or $(3,5,13)$.*

(ii) *If $\omega(q) = 4$, then*

$$(p_1, p_2, p_3, p_4) = (3,5,7,p_4); (3,5,11,p_4); (3,5,13,p_4); (3,5,17,p_4) \textit{ with } p_4 \le 251; (3,5,19,p_4) \textit{ with } p_4 \le 89; (3,5,23,p_4) \textit{with } p_4 \le 47; (3,5,29,p_4) \textit{ with } p_4 \le 31; (3,7,11,p_4) \textit{ with } p_4 \le 23; (3,7,13,p_4) \textit{ with } p_4 \le 19.$$

Proof: (i) Suppose $\omega(q) = 3$. Then we have , by (14) and (13) with $\beta = \frac{1}{2}$ that

$$p_1 < \frac{1}{1-(1/2)^{1/3}}$$

implying $p_1 = 3$. Next,

$$p_2 < \frac{1}{1-(3/4)^{1/2}}$$

giving $p_2 = 5, 7$. When $p_2 = 7$, we check that $2\varphi(q) > q$. Thus $p_2 = 5$. Lastly, we have

$$p_3 < \frac{1}{1-(15/16)}$$

giving $p_3 < 16$ i.e $p_3 = 7, 11, 13$, which proves the assertion.

Proof of (ii) is similar. □

3 Proofs of Theorems 1, 2 and Corollary

Proof of Theorem 2: Suppose S vanishes. Then Lemma 1 holds. Hence assertion (ii) of Lemma 1 is valid. On the other hand since (4) holds, we see that (5) is valid for every i with $1 \le i \le r$. In that case, by Lemma 4, assertion (ii) of Lemma 1 does not hold, a contradiction. □

Proof of Theorem 1: Let $\omega(q) = 3$. By the result of Okada, we assume that $2\varphi(q) \le q$. We consider those q for which

$$\frac{1}{2}\left(1-\frac{1}{h}\right) \le \frac{\varphi(q)}{q} \le \frac{1}{2}.$$

We first apply Lemma 5 (i) to get

$$(p_1, p_2, p_3) \in \{(3,5,7), (3,5,11), (3,5,13)\}.$$

Now we apply (13) with $\alpha = \frac{1}{2}(1-\frac{1}{h})$. We explain with an example. Let $(p_1, p_2, p_3) = (3,5,7)$. Then from (13) with $\alpha = \frac{1}{2}(1-\frac{1}{h})$ and (14) we have $\frac{1}{2}(1-\frac{1}{h}) \le \frac{48}{105}$ implying

$h \le 11$. Thus the only possible values of q are $3 \cdot 5 \cdot 7$ or $3^2 \cdot 5 \cdot 7$. By similar argument for the other values of (p_1, p_2, p_3) we obtain that

$$q \in \{3.5.7, 3^2.5.7, 3.5.11, 3.5.13, 3.5^2.11, 3.5^2.13, 3^2.5.11, \\ 3^2.5.13, 3^2.5^2.11, 3^2.5^2.13, 3^3.5.11, 3^3.5.13, 3^3.5^2.11, 3^3.5^2.13\}. \tag{15}$$

Thus for the values of q in (15) we find by Theorem 2 that S does not vanish. This proves the assertion of Theorem 1 for $\omega(q) = 3$.

Let $\omega(q) = 4$. We use Lemma 5 (ii) and (13) with $\alpha = \frac{1}{2}(1 - \frac{1}{h})$ to get an upper bound for α_i for every p_i. The upper bounds for α_i' s with $1 \le i \le 3$ are shown in the Table 1 for various values of p_1, p_2, p_3, p_4. Table 1 is read as follows. We always take $\alpha_4 = 1$. Suppose $(p_1, p_2, p_3, p_4) = (3, 5, 17, 251)$, then $\alpha_1 \le 8, \alpha_2 \le 5$ and $\alpha_3 \le 3$. Thus $3^8 \cdot 5^5 \cdot 17^3 \cdot 251$ is a value of q from the table. We see that S does not vanish for each of these q's by Theorem 2. □

Proof of Corollary. Let $q \le 1154$ and $q \notin \{525, 735, 945\}$. The only values of q for which (2) is not satisfied are $3.5.7, 3.5.11, 3.5.13, 3^2.5.7, 3^2.5.11, 3^2 \cdot 5 \cdot 13, 3 \cdot 5^2 \cdot 11, 3 \cdot 5^2 \cdot 13$ and S does not vanish for these values of q by Theorem 1(i). □

Table 1

(p_1, p_2, p_3)	p_4	α_1	α_2	α_3	(p_1, p_2, p_3)	p_4	α_1	α_2	α_3
(3,5,17)	19	2	1	1	(3,5,19)	71 – 73	5	3	1
	23	2	2	1		79 – 83	5	3	2
	29 – 61	3	2	1		89	6	4	2
	67 – 83	4	2	1	(3,5,23)	29 – 31	3	2	1
	89 – 113	4	3	1		37 – 41	4	3	1
	127 – 131	5	3	1		43	5	3	1
	137 – 181	5	3	2		47	5	4	2
	191 – 229	6	4	2	(3,5,29)	31	5	3	1
	233	7	4	2	(3,7,11)	13	2	1	1
	239 – 241	7	5	2		17	3	1	1
	251	8	5	3		19	3	2	1
(3,5,19)	23 – 43	3	2	1		23	4	2	2
	47 – 53	4	2	1	(3,7,13)	17	4	2	1
	59 – 67	4	3	1		19	6	3	2

References

[1] S.D. Adhikari, N. Saradha, T.N. Shorey and R. Tijdeman, Transcendental infinite sums, *Indag. Math.* (N.S) 12 (2001), 1-14.

[2] S.D. Adhikari, Transcendental Infinite sums and related questions, *Number Theory and discrete mathematics*, Proceedings of conference, Chandigarh, 2000 (Hindustan Book Agency, 2002), 169-178.

[3] S.D. Adhikari and N. Saradha, Arithmetic nature of sums of certain convergent series, *Resonance*, Vol 7, No.11 (2002), 35-46.

[4] A. Baker, B.J. Birch and E.A. Wirsing, On a problem of Chowla, *J. Number Theory* 5 (1973), 224-236.

[5] T. Okada, On certain infinite sums for periodic arithmetical functions, *Acta Arith.* 40 (1982), 143-153.

[6] N. Saradha and R. Tijdeman, On the transcendence of infinite sums of values of rational functions, *J. London Math. Soc.* 67 (2003), 1-13.

[7] R. Tijdeman, Some applications of diophantine approximations, *Number theory for the millennium*, Proceedings of conference, Urbana, IL, 2000-Vol. III (A.K. Peters, Natick, MA, 2002), 261-284.

N. Saradha
School of Mathematics,
Tata Institute of Fundamental Research,
Homi Bhabha Road, Colaba,
Mumbai 400 005, India.

e-mail: saradha@math.tifr.res.in

The Riemann Zeta Function and Related Themes – 2006, pp. 131–140

Powers in arithmetic progressions (III)

T.N. Shorey

Dedicated to Professor K. Ramachandra on his 70th birthday

1 Powers in products of terms in arithmetic progression

This is in contiuation of [30] and [31]. For an integer $v > 1$, we denote by $P(v)$ the greatest prime factor of v and $\omega(v)$ the number of distinct prime divisors of v. Further we put $P(1) = 1$ and $\omega(1) = 0$. The letter p always denotes a positive prime number. Let $b \geq 1, d \geq 1, k \geq 3, \ell \geq 2, n \geq 1$ and $y \geq 1$ be integers such that $P(b) \leq k, b$ is ℓ-free and $\gcd(n, d) = 1$. Let $\delta = 0, 1$ according as $P(b) < k$ or $P(b) = k$, respectively. We consider the equation

$$\Delta = \Delta(n, k, d) := n(n + d) \dots (n + (k - 1)d) = by^{\ell}. \tag{1}$$

We give an account on (1) and its extensions and prove a new result in section 2.

Let p divide two distinct terms $n + id$ and $n + jd$ of Δ. Then p divides $(i - j)d$. Thus p divides $i - j$ since $p \nmid d$ by $\gcd(n, d) = 1$. Thus $p \leq| i - j |< k$. Therefore a prime $p \geq k$ divides at most one term $n + i_p d$ of Δ. Now we count the power of $p \geq k + \delta$ on both the sides of (1) to conclude that $\text{ord}_p(n + i_p d) \equiv \text{ord}_p(\Delta) \equiv 0 \pmod{\ell}$ since $P(b) < k + \delta$. Therefore (1) admits the following factorisation

$$n + id = a_i y_i^{\ell} \text{ for } 0 \leq i < k$$

where $P(a_i) < k + \delta$ and a_i are ℓ-free. We observe that the assumption $P(b) < k + \delta$ ensures that there is no restriction on the factorisation of primes less than $k + \delta$ in Δ. It is natural to consider (1) with $P(b) < k$. In view of the binomial equation

$$\Delta(n, k, 1) = k! y^{\ell}$$

which has been completely solved by Erdös [8] for $k \geq 4$ and Györy [11] for $k = 3$, equation (1) is also considered when $P(b) = k$.

First we consider the case of consecutive positive integers i.e. of $d = 1$. Let $n > 1$ and $k = n! + 2 - \delta$. Then

$$\Delta(n - 1, k, 1) = (n - 1) \dots (n! + 1)(n! + 2) \dots (n! + n - \delta).$$

Therefore $P(\Delta(n - 1, k, 1)) \leq n! + 1$ since $n! + 2, \dots, (n! + n - \delta)$ are all composites. By writing $\Delta(n - 1, k, 1) = \Delta(n - 1, k, 1)1^{\ell}$, we observe that (1) has infinitely many solutions. Therefore we suppose that $P(\Delta) \geq k + \delta$. Then Erdös and Selfridge [10] proved that (1) with $P(b) < k$ does not hold. Furthermore, they showed that the assumption $P(\Delta) \geq k$ is not required in the case $b = 1$. Thus a product of two or more consecutive positive integers is never a power. If $P(b) = k$, Saradha [23] for $k \geq 4$ and Györy [12] for $k = 3$ showed that (1) is possible only when $(b, k, n) = (6, 3, 48)$.

Now we suppose that $d > 1$ throughout in §1. The first result is due to Euler [see [18]) that a product of four terms in arithmetic progression is never a square. This is equivalent to showing that (1) with $k = 4, \ell = 2$ and $b = 1$ does not hold by re-writing (1) as

$$n_1(n_1 + d_1)(n_1 + 2d_1)(n_1 + 3d_1) = b(\frac{y}{\mu})^2 \text{ with } \mu = \gcd(n, d), n_1 = n/\mu, d_1 = d/\mu.$$

Let $\ell > 2$ be prime. Assume that $n = y_0^\ell, n + d = y_1^\ell, n + 2d = y_2^\ell$ are three ℓ-th powers in arithmetic progression. Then $2y_1^\ell = y_0^\ell + y_2^\ell$ implying $y_0 = y_1 = y_2 = 1$ by Darmon and Merel [6]. This is a contradiction. Next we obtain an extension of the above result by showing that

$$n(n + d)(n + 2d) = y^\ell \text{ with } \ell > 2$$

is not possible. We have $n + id = a_i y_i^\ell$ with $P(a_i) \leq 2$. The possibility that $a_0 = a_1 = a_2 = 1$ is already excluded. Therefore we have either

$$n = 2y_0^\ell, n + d = y_1^\ell, n + 2d = 2^{\ell-1} y_2^{\ell_2}$$

or

$$n = 2^{\ell-1} y_0^\ell, n + d = y_1^\ell, n + 2d = 2y_2^\ell$$

The latter possibility is the mirror image of the former and it suffices to confine to the former. Then

$$2y_1^\ell = 2y_0^\ell + 2^{\ell-1} y_2^\ell$$

i.e.

$$y_1^\ell = y_0^\ell + 2^{\ell-2} y_2^\ell.$$

This is not possible by Ribet [22]. This proves the assertion and it was due to Györy [13].

Now we consider an analogue of Euler's theorem for higher powers. For this, we consider

$$n(n + d)(n + 2d)(n + 3d) = y^\ell \text{ with } \ell > 2. \tag{2}$$

We have $n + id = a_i y_i^\ell$ with $P(a_i) \leq 3$. We may assume that $P(a_i) = 3$ otherwise the assertion follows as above. We give a proof when d is even and the proof for the case d odd is similar. Then n is odd since $\gcd(n, d) = 1$ and all the a_i's are odd. We may suppose that at least one a_i is divisible by 3 otherwise they are all equal to 1 and this possibility is already excluded. Since a_i are ℓ-free, we observe that a_0 and a_3 are divisible by 3. Thus either

$$n = 3y_0^\ell, n + d = y_1^\ell, n + 2d = y_2^\ell, n + 3d = 3^{\ell-1} y_3^\ell \tag{3}$$

or its mirror image. It suffices to restrict to (3). Now we use the relation

$$(n + d)(n + 2d) - n(n + 3d) = 2d^2$$

i.e.

$$(y_1, y_2)^\ell - (3y_0 y_3)^\ell = 2d^2$$

which is not possible when $\ell \geq 5$ and ℓ prime by Bennett and Skinner [5]. Further the case $\ell = 3$ of (2) has been excluded by using old results of Selmer [27], see also [4] for a correction. Thus (2) is not possible extending the result of Euler for higher powers. This has been proved by Győry, Hajdu and Saradha [14] which also contains an analogous result for $k = 5$. Further Bennett, Bruin, Győry and Hajdu [4] showed that the contributions on Fermat's equation led to proving that (1) with $6 \leq k \leq 11, \ell > 2$ and $b = 1$ is not possible. If $\ell = 2$, Obláth [21] for $k = 5$ and Hirata-Kohno, Laishram, Shorey and Tijdeman [16] for $6 \leq k \leq 109$ proved that (1) with $b = 1$ is not possible. A weaker version of the preceding result with $6 \leq k \leq 11$ was proved independently, by Bennett, Bruin, Győry and Hajdu [4]. If $\ell > 2$, and $6 \leq k \leq 11$ the assumption $b = 1$ has been relaxed in [4] to $P(b) \leq 5$. For $\ell = 2$, the assumption $b = 1$ has been relaxed to $P(b) < k$ by Mukhopadhyay and Shorey

[18] for $k = 5$, Bennett, Bruin, Györy and Hajdu [4] for $k = 6$ and Hirata-Kohno, Laishram, Shorey and Tijdeman [16] for $7 \leq k \leq 101$. Finally, we remark that Bennett, Bruin, Györy and Hajdu [4] proved, independently, a version of the preceding result when $7 \leq k \leq 11$ and $P(b) \leq 5$.

We have considered (1) so far with fixed k. It is clear that the above approach is not suitable for (1) with k as variable. Now we turn to considering (1) with k as variable. We begin with the following

Conjecture: Equation (1) implies that

$$(k, \ell) \in \{(4, 2), (3, 2), (3, 3)\}.$$

It is necessary to exclude the above three possibilities as in each of these cases, we can find at least one value of b when (1) has infinitely many solutions. It is a very difficult conjecture. We consider an easier question whether there are infinitely many d for which (1) does not hold. This has been solved. Shorey and Tijdeman [32] proved that (1) implies that k is bounded by an effectively computable number depending only on ℓ and $\omega(d)$. In particular, we see that k is bounded by an absolute constant whenever (1) with $\ell = 2$ and $\omega(d) = 1$ holds. In fact, it follows from the results of Saradha and Shorey [25] for $k \geq 9$ and Mukhopadhyay and Shorey [18] for $4 \leq k \leq 8$ that (1) with $\ell = 2, P(b) < k$ and $\omega(d) = 1$ is possible only when $(b, d, k, n, y) = (6, 23, 4, 75, 4620)$. Further the assumption $P(b) < k$ has been relaxed to $P(b) \leq k$ in [25] and [18] except for the case $k = 5, P(b) = 5$. Also the assumption $\gcd(n, d) = 1$ is not required when $b = 1$. Thus these results include a theorem of Saradha and Shorey [25] that a product of four or more terms in arithmetic progression with common difference a prime power is never a square. Further it has been shown in [30] that (1) with $k = 3$ and $b = 1$ holds if and only if $d^2 - 2Y^2 = -1$ where $Y > 0$ is an integer. This equation has infinitely many solutions in integers $d > 0$ and $Y > 0$. On the other hand, we do not know whether it has finitely or infinitely many solutions in integers $d > 0$ and $Y > 0$ such that d is prime. Thus the case $k = 3$ of the preceding result of Saradha and Shorey remains open. Finally Laishram and Shorey [17] proved that (1) with $\ell = 2, k \geq 4, P(b) < k$ and $\omega(d) = 2, 3, 4$ is not possible.

It is well-known that (1) with $k = 4, \ell = 2$ and $b = 6$ has infinitely many solutions for general d. On the other hand, it has only one $n = 75, d = 23, y = 4620$ when d is restricted to prime powers. The proof depends on the theory of linear forms in logarithms via the method of Baker-Davenport [1] on solving a pair of Pell's equations simultaneously. We give a proof. Assume that

$$n(n + d)(n + 2d)(n + 3d) = 6y^2 \text{ with } \omega(d) = 1.$$

Then $n + id = a_i y_i^2$ with $P(a_i) \leq 3$. Put $R = \{a_0, a_1, a_2, a_3\}$. Then $R \subseteq \{1, 2, 3, 6\}$. We observe that d is odd by $\gcd(n, d) = 1$ and $| R |\neq 1, 4$ by the result of Euler. If $| R |= 2$, we again see from the result of Euler that three a_i's are equal to 1 and the fourth one is 6. Thus at least two $n + id$ and $n + jd$ with $i \neq j$ are odd square. Therefore $(i - j)d \equiv 0 \pmod 8$ implying that d is even, a contradiction. Therefore $| R |= 3$ and 3 divides exactly one a_i. Let $3 \mid a_0$. Then we observe that

$$n = 3y_0^2, n + d = 2y_1^2, n + 2d = y_2^2, n + 3d = y_3^2.$$

By subtracting the fourth equation from the third, we get $d = y_3^2 - y_2^2$ implying $y_3 - y_2 = 1, y_3 + y_2 = d$ since d is an odd prime power. Thus $d = 2y_2 + 1$. Now $n = n + 2d - 2d =$

$y_2^2 - 4y_2 - 2 = (y_2 - 2)^2 - 6$ and $n + d = n + 2d - d = y_2^2 - 2y_2 - 1 = (y_2 - 1)^2 - 2$. Therefore

$$3y_0^2 = (y_2 - 2)^2 - 6, 2y_1^2 = (y_2 - 1)^2 - 2.$$

Now we conclude from Baker-Davenport method already referred that the above pair of equations is possible only when $y_0 = 5, y_1 = 7, y_2 = 11$ implying $n = 75, d = 23, y = 4620$. Baker-Davenport method can be applied in more general context when (1) holds with k fixed, $\omega(d) = 1$ and $a_i = a_j$ for $i \neq j$. If we find two a_i's equal to 1 and three a_i's equal to 2,3 and 6, then we can apply Runge's method as in [25], [18] and [19]. We shall apply these two methods in the next section.

Next we take $\ell > 2$ with ℓ prime and we continue considering our question of finding infinitely many d such that (1) does not hold. Saradha and Shorey proved that (1) with $k \geq 4$ and $P(d) \leq 5$ does not hold. This is a consequence of a theorem of Saradha and Shorey [24] that (1) with $k \geq 4$ implies that d is divisible by a prime congruent to $1 \pmod{\ell}$. There exists a prime p dividing d such that $p \geq 1 + 2\ell \geq 7$ implying the above assertion. We observe that there are infinitely many d such that $P(d) \leq 5$. The case $k = 3$ remains open. This completes our examination of the question under consideration.

The Conjecture has been confirmed for a large number of values of d. Let $k \geq 4$ if $\ell > 3$. Then Saradha and Shorey [26] proved that (1) implies

$$d > d'$$

where

$$d' = \begin{cases} 104 & \text{if } \ell = 2 \\ 30 & \text{if } \ell = 3 \\ 950 & \text{if } \ell = 4 \\ 5.10^4 & \text{if } \ell = 5, 6 \\ 10^8 & \text{if } \ell = 7, 8, 9, 10 \\ 10^{15} & \text{if } \ell \geq 11. \end{cases}$$

Further the preceding estimate for $\ell = 2$ has been sharpened considerably in Laishram and Shorey [17].

2 Powers in products of terms in arithmetic progression with one term omitted

We consider (1) with one term omitted on the left hand side. Erdős and Selfridge [10] conjectured that

$$\frac{6!}{5} = (12)^2, \ \frac{10!}{7} = (720)^2, \ \frac{4!}{3} = 2^3$$

are the only powers which are products of $k - 1$ integers out of $k (\geq 3)$ consecutive positive integers. This has been confirmed by Saradha and Shorey [25]. Now we consider a more general equation than the underlying one in the above result:

$$n(n+1)\dots(n+i_0-1)(n+i_0+1)\dots(n+k-1) = by^\ell \quad (4)$$

where $0 < i_0 < k, P(b) \leq k$ and b is ℓ-free. We suppose that the left hand side of (4) is divisible by a prime exceeding k. Then we have: Let $k \geq 3$ and $k \geq 4$ if $\ell = 2$. Then (4) is possible only when $n = 24, k = 4, i = 2, b = 2, y = 90$. This was proved by Hanrot, Saradha and Shorey [15] for $k \geq 8, \ell > 2$, Bennett [3] for $3 \leq k \leq 7, \ell > 2$ and Saradha and Shorey [25] for $\ell = 2$. Next we consider an analogue of (4) with $\ell = 2$ for arithmetic progressions:

$$n(n+d)\dots(n+(i-1)d)(n+(i+1)d)\dots(n+(k-1)d) = by^2 \tag{5}$$

where $0 < i_0 < k, d > 1, P(b) < k$ and b is square free. We suppose that $k \geq 5$ since (5) has infinitely many solutions when $k = 3, 4$. It has been proved by Saradha and Shorey [25] for $k \geq 30$ and Mukhopadhyay and Shorey [19] for $k \geq 9$ that (5) with $\omega(d) = 1$ does not hold. Thus it remains to consider (5) with $\omega(d) = 1$ for $k \leq 8$. If $b = 1$, we prove:

Theorem. *Equation* (5) *with* $k \in \{6, 7, 8\}, \omega(d) = 1$ *and* $b = 1$ *does not hold.*

Thus (5) with $k \geq 6, \omega(d) = 1$ and $b = 1$ is not possible. The case $k = 5$ remains open. For general d, it is known that (5) with $b = 1$ has infinitely many solutions, see Dickson [7, p. 440].

Proof: We shall always assume that $b = 1$ in the proof. By (5), we have

$$n + id = a_i y_i^2 \text{ for } 0 \leq i < k \text{ and } i \neq i_0$$

where $P(a_i) < k$ and a_i is square free. Since $b = 1$ and a_i is square free, we see from (5) that if a prime p divides a_i, then p divides a_j for some $j \neq i$. This observation will be used several times in the following proof. We put $R' = \{a_0, a_1, \dots, a_{i_0-1}, a_{i_0+1}, \dots, a_{k-1}\}$ and we write T for the set of all $a_i \in R'$ composed of 2 and 3. We observe that $T \subset \{1, 2, 3, 6\}$. As in [18] and [19], we may suppose that $|R'| \geq k-2$.

Let $k = 8$. First we show that 5 divides a_i and 7 divides a_j for some i and j. If $|R'| = k-1$ and none of the a_i is divisible by 5 or 7, then $|T| \geq k-3 = 5$, a contradiction. Thus we may suppose that $|R'| = k-2$. Assume that none of the a_i is divisible by 7. Then 5 divides a_i for some i otherwise $|T| = |R'| = k-2 = 6$, a contradiction. Thus 5 divides a_0, a_5 or a_1, a_6 or a_2, a_7. By considering the mirror image of (5), we may suppose that 5 divides a_0, a_5 or a_1, a_6. Let 5 divide a_0, a_5. Then, by using Legendre symbol $\pmod 5$ as in [18] and [19], we derive that $|T| = 4$ such that

$$\{a_1, a_4, a_6\} \cap T \subseteq \{1, 6\}, \{a_2, a_3, a_7\} \cap T \subseteq \{2, 3\} \tag{6}$$

or

$$\{a_1, a_4, a_6\} \cap T \subseteq \{2, 3\}, \{a_2, a_3, a_7\} \cap T \subseteq \{1, 6\}. \tag{7}$$

We may suppose that a_6 or a_7 is not the omitted term otherwise a_1, a_2, a_3, $a_4 \in T$ and the assertion follows from the result in [18] for $k = 4$ stated in section 1. First we consider (6). Let a_1 be the omitted term. Then $a_4 = 6$ or $a_6 = 6$. Therefore a_3 and a_7 are $\equiv 3 \pmod 8$ implying d is even, a contradiction. Let a_4 be the omitted term. Then $a_6 = 6$ or $a_1 = 6$. The possibility $a_6 = 6$ is excluded as above whereas $a_1 = 6$ implies that $a_7 = 3$ and $a_2 = a_3 = 2$, a contradiction. If a_2 is the omitted term, then $a_3 = 2$ or $a_7 = 2$ implying that a_4 and a_6 are odd squares, a contradiction. Let a_3 be the omitted term. Then $a_2 = 2$ or $a_7 = 2$. The case $a_7 = 2$ is already excluded. If $a_2 = 2$, then $a_7 = 3, a_4 = 6$ and $a_1 = a_6 = 1$. This is excluded by Runge's method as in [18] and [19]. Further the possibility (7) is excluded similarly. The proof for the case 5 dividing a_1, a_6 is similar. □

Now we suppose that none of a_i is divisible by 5. Then 7 divides a_0, a_7 and, by using Legendre symbol mod 7, we have

$$\{a_1, a_2, a_4\} \cap T \subseteq \{1, 2\}, \{a_3, a_5, a_6\} \cap T \subseteq \{3, 6\} \tag{8}$$

or

$$\{a_1, a_2, a_4\} \cap T \subseteq \{3, 6\}, \{a_3, a_5, a_6\} \cap T \subseteq \{1, 2\}. \tag{9}$$

First we consider (8). We observe that a_5 is the omitted term. If $a_3 = 6$, then a_2 and a_4 are odd squares, a contradiction. If $a_6 = 6$, then $a_3 = 3$, $a_2 = 2$ and $a_1 = a_4 = 1$ by mod 3. This possibility is excluded again by Runge's method. The proof that (9) does not hold is similar.

Thus $5 \mid a_i$ and $7 \mid a_j$ for some i and j. Then 7 divides a_0, a_7. Further we may suppose that 5 divides a_0, a_5 or a_1, a_6. If $|R'| = k - 2$, then $a_i = a_j$ for some $i \neq j$ and we exclude this case by Baker-Davenport method or Runge's method as in [18] and [19]. Thus we may suppose that $|R'| = k - 1$. Let 5 divide a_0, a_5. Then $T = \{1, 2, 3, 6\}$. Since 1.2.3.6 is a square, we see from (5) that $a_0 a_5 a_7$ is a square. Therefore $a_0 = 35$, $a_5 = 5$, $a_7 = 7$ or $a_0 = 35$, $a_5 = 10$, $a_7 = 14$. Further, by mod 7, we have

$$\{a_1, a_2, a_4\} \cap T \subseteq \{1, 2\}, \{a_3, a_6\} \cap T \subseteq \{3, 6\} \tag{10}$$

or

$$\{a_1, a_2, a_4\} \cap T \subseteq \{3, 6\}, \{a_3, a_6\} \cap T \subseteq \{1, 2\}. \tag{11}$$

Assume (11). Then a_2 is the omitted term and a_1, a_4 are distinct elements of T such that $\{a_1, a_4\} = \{3, 6\}$. This is not possible since $(\frac{a_1}{5}) = (\frac{a_4}{5})$ and $(\frac{3}{5}) \neq (\frac{6}{5})$. Now we suppose (10). We observe by mod 3 that a_2 is not the omitted term. Let $a_3 = 6$. Then $a_6 = 3$ and $n + 6d \equiv 3 \mod 8$. Also $n \equiv 35 \pmod 8$. Therefore $6d \equiv 0 \pmod 8$ which is not possible since d is odd. Thus we may suppose that $a_6 = 6$. Then

$$a_0 = 35, a_1 = 1, a_2 = 2, a_3 = 3, a_5 = 5, a_6 = 6, a_7 = 7 \text{ with } a_4 \text{ omitted} \tag{12}$$

or

$$a_0 = 35, a_2 = 2, a_3 = 3, a_4 = 1, a_5 = 5, a_6 = 6, a_7 = 7 \text{ with } a_1 \text{ omitted.} \tag{13}$$

First we consider (13). We use the identity $3(n + 2d) = n + 2(n + 3d)$ to obtain

$$6y_2^2 - 6y_3^2 = 35y_0^2$$

where y_2 and y_3 are odd. Thus

$$\frac{y_2 + y_3}{2} \frac{y_2 - y_3}{2} = 210z_0^2$$

for some positive integers z_0. Therefore

$$y_2 = \ell\zeta^2 + m\eta^2, y_3 = \ell\zeta^2 - m\eta^2$$

where ℓ, m, ζ and η are positive integers such that $\gcd(\ell\zeta, m\eta) = 1$ and $\ell m = 210$. Now we have

$$d = (n + 3d) - (n + 2d) = 3(\ell\zeta^2 - m\eta^2)^2 - 2(\ell\zeta^2 + m\zeta^2)^2$$

i.e.

$$d = \ell^2\zeta^4 + m^2\eta^4 - 10\ell m\zeta^2\eta^2.$$

Now we turn to (12). By using $2(n + d) = n + n + 2d$, we obtain similarly

$$y_6^2 = \ell^2\zeta^4 + m^2\eta^4 - \frac{14}{3}\ell m\zeta^2\eta^2$$

where ℓ, m, ζ and η are positive integers such that $\gcd(\ell\zeta, m\eta) = 1$ and $\ell m = 70$. Now we consider

$$Y^2 = \ell^2\zeta^4 + m^2\eta^4 - \phi\ell m\zeta^2\eta^2$$

where $\phi = 46/7$ if $\ell m = 210$ and $\phi = 14/3$ if $\ell m = 70$. Let $X_1 = \zeta/\eta$ and $Z_1 = Y/\eta^2$. Then

$$Z_1^2 = \ell^2X_1^4 - \phi\ell mX_1^2 + m^2.$$

Let X_2 be given by

$$Z_1 = -\ell X_1^2 + 2X_2 + \frac{\phi m}{6}.$$

Now we put

$$Z_2 = 2\ell X_1\left(X_2 - \frac{\phi m}{6}\right)$$

and

$$Z = 864\ell Z_2, X = 144\ell X_2.$$

Then we derive the elliptic curve

$$Z^2 = X^3 - 432(\phi^2 + 12)(\ell m)^2X - 3456\phi(\phi^2 - 36)(\ell m)^3$$

over Q. The above equation is satisfied for $X = X_1$ and $Z = Z_1$ with $X_1, Z_1 \in Q$ and $Z_1 \neq 0$. Now we check by MAGMA that the rank of the elliptic curve is zero and it has no torsion point with $Z \neq 0$. This is a contraction excluding the possibilities (12) and (13).

Let 5 divide a_1, a_6. By mod 7, we obtain

$$\{a_2, a_4\} \cap T \subseteq \{1, 2\}, \{a_3, a_5\} \cap T \subseteq \{3, 6\} \tag{14}$$

or

$$\{a_2, a_4\} \cap T \subseteq \{3, 6\}, \{a_3, a_5\} \cap T \subseteq \{1, 2\}. \tag{15}$$

First we suppose (14). Then a_5 is the omitted term otherwise 3 divides a_5 and 3 divides no other a_i. Thus $a_3 = 3$ or 6. If $a_3 = 6$, then a_2 and a_4 are odd squares, a contradiction. Thus $a_3 = 3$. Consequently $a_2 = 1$, $a_3 = 3$ and $a_4 = 2$ by mod 5. Thus for $\delta_1, \delta_2 \in \{0, 1\}$, we have

$$a_0 = 2^{\delta_1}3^{\delta_2}7, a_1 = 5, a_2 = 1, a_3 = 3, a_4 = 2, a_6 = 2^{1-\delta_1}3^{1-\delta_2}5, a_7 = 7. \tag{16}$$

Similarly we obtain the mirror image of (16) when (15) holds. Thus it suffices to exclude (16). We observe that $(\delta_1, \delta_2) \neq (0, 0), (1, 1)$ since $| R' |= k - 1$. Further we apply (36) or (34) with $b = 6$ or 10 of [4, Lemma 4.1] if $(\delta_1, \delta_2) = (0, 1)$ or $(\delta_1, \delta_2) = (1, 0)$, respectively, to exclude (16). This completes the proof for the case $k = 8$.

Let $k = 7$. Then $| R' |\geq 5$. Further we may suppose that $| R' |= k - 1 = 6$ otherwise the assertion follows from Baker-Davenport method. Then 5 divides a_0, a_5 or a_1, a_6. By considering the mirror image of (5) with $b = 1$, we may suppose that 5 divides a_0, a_5.

Observe that $a_0 \neq a_5$ and $| T |= 4$ so that all the elements of T are distinct. Therefore a_0, a_5 is a square. This is not possible as a_0 and a_5 are square free.

Let $k = 6$. Then $| R' | \geq 4$ and we may suppose that $| R' |= 5$ by Baker-Davenport method. Then 5 divides a_0, a_5 and

$$\{a_1, a_4\} \cap T \subseteq \{1, 6\}, \{a_2, a_3\} \cap T \subseteq \{2, 3\} \tag{17}$$

or

$$\{a_1, a_4\} \cap T \subset \{2, 3\}, \{a_2, a_3\} \cap T \subseteq \{1, 6\}. \tag{18}$$

First we assume (17). Then no a_i equals 6. Therefore $a_0 = 30$, $a_1 = 1$, $a_2 = 2$, $a_3 = 3$, $a_5 = 5$ or $a_0 = 5$, $a_1 = 1$, $a_2 = 3$, $a_3 = 2$, $a_5 = 30$ or $a_0 = 30$, $a_2 = 2$, $a_3 = 3$, $a_4 = 1$, $a_5 = 5$ or $a_0 = 5$, $a_2 = 3$, $a_3 = 2$, $a_4 = 1$, $a_5 = 30$. These possibilities are excluded by (36) of [4, Lemma 4.1]. The proof for excluding (18) is similar. This completes the proof of the Theorem.

3 Powers in products of integers from a block of consecutive integers

We turn to a more general question than considered in sections 1 and 2 for consecutive integers. Let $d_1 < d_2 \ldots < d_t$ be integers in $[0, k)$, $P(b) \leq k$ and

$$(n + d_1) \ldots (n + d_t) = by^\ell. \tag{19}$$

We assume that the left hand side of (19) is divisible by a prime exceeding k. Let $t = k$, then $d_i = i$ and (19) is (1) with $d = 1$. If $t = k - 1$, then (19) and (4) are identical. Let $\ell > 2$. Shorey [28], [29] showed that (19) with $t \geq \frac{47}{56}k$ implies that k is bounded by an effectively computable absolute constant. Further Nesterenko and Shorey [20] replaced $\frac{47}{56}$ by 49 if $\ell \geq 7$. The proofs depend on the theory of linear forms in logarithms with $\alpha_i' s$ close to 1, the estimates of Baker on the approximations of certain algebraic numbers by rationals based on hypergeometric method and the method of Roth and Halberstam on difference between consecutive ν-free integers. If $\ell = 2$, Shorey [29] showed that we can take

$$t \geq k\left(1 - \frac{(1 - \epsilon)\log\log k}{\log k}\right) \text{ for } k \geq k_0(\epsilon). \tag{20}$$

This answers a question of Erdős [9]. The proof depends on a theorem of Siegel and Baker that hyperelliptic equation, under necessary assumption, has only finitely many integer solutions. The assumption (20) has been relaxed in [2]. An analogue of (19) for arithmetic progressions has been considered in [33].

References

[1] A. Baker and H. Davenport, The equations $3x^2 - 2 = y^2$ and $8x^2 - 7 = z^2$, *Quart. Jour. Math.* Oxford Ser. (2) **20** (1969), 129–137.

[2] R. Balasubramanian and T.N. Shorey, Squares in products from a block of consecutive integers, *Acta Arith.* **65** (1994), 213–220.

[3] M.A. Bennett, Products of consecutive integers, *Bulletin London Math. Soc.* (2004).

[4] M.A. Bennett, N. Bruin, K. Győry and L. Hajdu, Powers from products of consecutive terms in arithmetic progression, *Proc. London Math. Soc.*, to appear.

[5] M.A. Bennett and C. Skinner, Ternary Diophantine equations via Galois representations and modular forms, *Canadian Jour. Math.* **56** (2004), 23–54.

[6] H. Darmon and L. Merel, Winding quotients and some variants of Fermat's Last Theorem, *Jour. Reine Angew. Math.* **490** (1997), 81–100.

[7] L.E. Dickson, *History of the Theory of Numbers*, Vol.II, Chelsea Publ. Co. (1952).

[8] P. Erdős, On a diophantine equation, *Jour. London Math. Soc.* **26** (1951), 176–178.

[9] P. Erdős, On the product of consecutive integers III, *Indag. Math.* **17** (1955), 85–90.

[10] P. Erdős and J.L. Selfridge, The product of consecutive integers is never a power, *Illinois Jour. Math.* **19** (1975), 292–301.

[11] K. Győry, On the diophantine equation $\binom{n}{k} = x^\ell$, *Acta Arith.* **80** (1997), 289–295.

[12] K. Győry, On the diophantine equation $n(n+1)\dots(n+k-1) = bx^\ell$, *Acta Arith.* **83** (1998), 87–92.

[13] K. Győry, Power values of products of consecutive integers and binomial coefficients, *Number Theory and Its applications*, Kluwer Acad. Publ. (1999), 145–156.

[14] K. Győry, L. Hajdu and N. Saradha, On the Diophantine equation $n(n+d)\dots(n+(k-1)d) = by^\ell$, *Canadian Math. Bulletin* 47 (2004), 373–384.

[15] G. Hanrot, N. Saradha and T.N. Shorey, Almost perfect powers in consecutive integers, *Acta Arith.* **99** (2001), 13–25.

[16] Noriko Hirata-Kohno, S. Laishram, T.N. Shorey and R. Tijdeman, An extension of a theorem of Euler, to appear.

[17] S. Laishram and T.N. Shorey, The equation $n(n+d)\dots(n+(k-1)d) = by^2$ *with* $\omega(d) = 2, 3, 4, 5$, to appear.

[18] A. Mukhopadhyay and T.N. Shorey, Almost squares in arithmetic progression (II), *Acta Arith.* **110** (2003), 1–14.

[19] A. Mukhopadhyay and T.N. Shorey, Almost squares in arithmetic progression (III), *Indag. Math.* (2005).

[20] Yu. V. Nesternko and T.N. Shorey, Perfect powers in products of integers from a block of consecutive integers (II), *Acta Arith.* **76** (1996), 191–198.

[21] R. Obláth, Über das produckt funf aufeinander folgender zahlen in einer arithmetischen reihe, *Publ. Math. Debrecen* 1 (1950), 222–226.

[22] K.Ribet, On the equation $a^p + 2^\alpha b^p + c^p = 0$, *Acta Arith.* **79** (1997) 7–16.

[23] N. Saradha, On perfect powers in products with terms from arithmetic progressions, *Acta Arith.* **82** (1997), 147–172.

[24] N. Saradha and T.N. Shorey, Almost perfect powers in arithmetic progression, *Acta Arith.* **99** (2001), 363–388.

[25] N. Saradha and T.N. Shorey, Almost squares in arithmetic progression, *Compositio Math.* **138** (2003), 73–111.

[26] N. Saradha and T.N. Shorey, Contributions towards a conjecture of Erdős on perfect powers in arithmetic progressions, *Compositio Math.*, 141 (2005), 541–560.

[27] E. Selmer, The diophantine equation $ax^3 + by^3 + cz^3 = 0$, *Acta Math.* **85** (1951), 205–362.

[28] T.N. Shorey, Perfect powers in values of certain polynomials at integer points, *Math. Proc. Camb. Philos. Soc.* **99** (1986), 195–207.

[29] T.N. Shorey, Perfect powers in products of integers from a block of consecutive integers, *Acta Arith.* **49** (1987), 71–79.

[30] T.N. Shorey, Powers in arithmetic progression, In: G. Wüstholz (ed.) *A Panorama in Number Theory or The view from Baker's Garden*, Cambridge Univ. Press (2002), 325–336.

[31] T.N. Shorey, Powers in arithmetic progression (II), *Analytic Number Theory*, RIMS Kokyuroku (2002), Kyoto University.

[32] T.N. Shorey and R. Tijdeman, Perfect powers in products of terms in an arithmetical progression, *Compositio Math.* **75** (1990), 307–344.

[33] T.N. Shorey and R. Tijdeman, Perfect powers in products of terms in an arithmetical progression (III), *Acta Arith.* 61 (1992), 391–398.

T.N. Shorey
School of Mathematics,
Tata Institute of Fundamental Research,
Homi Bhabha Road, Colaba,
Mumbai 400 005, India.

e-mail: shorey@math.tifr.res.in

The Riemann Zeta Function and Related Themes – 2006, pp. 141–154

Maximal order of certain sums of powers of the logarithmic function

V. Sitaramaiah and M.V. Subbarao*†

Dedicated to Professor K. Ramachandra on his 70th birthday

1 Introduction.

In 1972, Erdös and Zaremba [2] proved that

$$\overline{\lim_{n\to\infty}}(\log\log n)^{-2}\sum_{d|n}\frac{\log d}{d}=e^{\gamma}, \tag{1}$$

where γ is Euler's constant.

In [5] we observed that there is a wide class of arithmetic functions for which results of the type given in (1) hold. We mention here two such examples from this class:

$$\overline{\lim_{n\to\infty}}(\log\log n)^{-2}\sum_{d|n}\frac{\log\sigma(d)}{d}=e^{\gamma},$$

and

$$\overline{\lim_{n\to\infty}}((\log\log n)(\log\log\log n))^{-1}\sum_{d|n}\frac{\Omega(d)}{d}=e^{\gamma},$$

where $\sigma(n)$ is the sum of the divisors of n and $\Omega(n)$ is the total number of prime factors of n.

If one thinks of generalizing (1), one way is to replace $\log d$ in (1) by $(\log d)^k$ where k is any positive integer and $\log\log n$ by a suitable power of $\log\log n$ to ensure that the limit sup of the corresponding function is finite and positive and ask what the right hand side would be. We believe that for each positive integer k,

$$\overline{\lim_{n\to\infty}}(\log\log n)^{-(k+1)}\sum_{d|n}\frac{(\log d)^k}{d}=c_k e^{\gamma}, \tag{2}$$

where c_k is a positive constant.

In this paper we could establish the validity of (2) when $k = 2$ or 3. In fact we prove that (see Theorems 1 and 2)

$$\overline{\lim_{n\to\infty}}(\log\log n)^{-3}\sum_{d|n}\frac{(\log d)^2}{d}=\frac{3}{2}e^{\gamma}, \tag{3}$$

and

$$\overline{\lim_{n\to\infty}}(\log\log n)^{-4}\sum_{d|n}\frac{(\log d)^3}{d}=\frac{17}{6}e^{\gamma}. \tag{4}$$

Our method of proof of (3) and (4) shows that the results (3) and (4) are independent. This appears to be the main difficulty in establishing (2) for any k.

*Partially supported by an NSERC grant.

†Editors regret to record the passing away of Prof. M. V. Subbarao on February 15, 2006.

2 Preliminaries

For each positive integer k,let

$$S_k(n) = \sum_{d|n} \frac{(\log d)^k}{d}, \tag{5}$$

and

$$B_k(n) = \frac{S_k(n)}{F(n)}, \tag{6}$$

where

$$F(n) = \sum_{d|n} \frac{1}{d} = \frac{\sigma(n)}{n}, \tag{7}$$

$\sigma(n)$ being the sum of the divisors of n. In the following we state Lemmas 1 to 3 without proofs: We have

Lemma 1. *When $(m\,, n) = 1$, we have*

$$B_k(mn) = B_k(m) + B_k(n) + \sum_{i=1}^{k-1} \binom{k}{i} B_i(m) B_{k-i}(n), \tag{8}$$

where as usual empty sums are assigned the value zero.

For later reference we record here the following special cases of lemma 1; when $(m\,, n) = 1$, we have

$$B_1(mn) = B_1(m) + B_1(n), \tag{9}$$

so that B_1 is an additive function.

$$B_2(mn) = B_2(m) + B_2(n) + 2B_1(m)B_1(n), \tag{10}$$

and

$$B_3(mn) = B_3(m) + B_3(n) + 3B_1(m)B_2(n) + 3B_1(n)B_2(m). \tag{11}$$

Lemma 2. *If $1 < n = p_1^{\alpha_1} p_2^{\alpha_2} \ldots p_r^{\alpha_r}$ is the canonical representation of n, then*

$$B_2(n) = \sum_{i=1}^{r} B_2(p_i^{\alpha_i}) + 2 \sum_{1<i<j\le r} B_1(p_i^{\alpha_i}) B_1(p_j^{\alpha_j}). \tag{12}$$

Lemma 3. *If $1 < n = p_1^{\alpha_1} p_2^{\alpha_2} \ldots p_r^{\alpha_r}$ is the canonical representation of n, then*

$$B_3(n) = \sum_{i=1}^{r} B_3(p_i^{\alpha_i}) + 3 \sum_{\substack{1\le i\le r \\ 1\le j\le r \\ i\ne j}} B_2(p_i^{\alpha_i}) B_1(p_j^{\alpha_j}) + 6 \sum_{1\le i<j<k\le r} B_1(p_i^{\alpha_i}) B_1(p_j^{\alpha_j}) B_1(p_k^{\alpha_k}).$$

Note: Throughout this paper the letter p with or without suffixes denotes a prime number.

Lemma 4. *For any positive integer k and $x \ge 2$,*

$$\sum_{p\le x} \frac{(\log p)^k}{p} = \frac{1}{k}(\log x)^k + O\left((\log x)^{k-1}\right).$$

Proof: We know (cf. [3], Theorem 4.25) that

$$\sum_{p\le x}\frac{\log p}{p}=\log x+O(1). \tag{13}$$

Now lemma 4 follows from (13), induction on k and partial summation.

Remark: The error term in (13) could be further improved. For example see [4]. However we will be requiring only the one given in lemma 4.

Lemma 5. *For any positive integer k and for large n, we have*

$$\sum_{p|n}\frac{(\log p)^k}{p}\le\frac{1}{k}(\log\log n)^k+O((\log\log n)^{k-1}(\log\log\log n)^k).$$

Proof: The proof is essentially the same as in the case $k=1$ given by Erdös and Zaremba [2]. However for the sake of completeness we give the details.

The function $(\log x)^k/k$ is decreasing for $x>N=N(k)$. Let $p_1,p_2,\dots,p_r$ denote the distinct prime factors of n with $p_1<p_2<\cdots<p_r$. We have

$$\begin{aligned}\sum_{p|n}\frac{(\log p)^k}{p} &= \sum_{\substack{p|n\\ p\le N}}\frac{(\log p)^k}{p}+\sum_{\substack{p|n\\ p>N}}\frac{(\log p)^k}{p}\\ &= O(1)+\sum_{\substack{p|n\\ p>N}}\frac{(\log p)^k}{p}\\ &= O(1)+\Sigma, \end{aligned}\tag{14}$$

say. We may assume that Σ is non-empty. Let s be the least positive integer such that $p_s>N$. Let q_i be the ith prime. We choose the first positive integer $t=t(N)$ such that $q_t>N$. We have

$$p_s\ge q_t\,,\,p_{s+1}\ge q_{t+1},\dots,p_r\ge q_{t+r-s}.$$

Therefore we have

$$\begin{aligned}\Sigma=\sum_{i=s}^{r}\frac{(\log p_i)^k}{p_i} &\le \sum_{i=t}^{t+r-s}\frac{(\log q_i)^k}{q_i}\\ &\le \sum_{i=1}^{r+t}\frac{(\log q_i)^k}{q_i}\\ &= \sum_{p\le q_{r+t}}\frac{(\log p)^k}{p}\\ &= \frac{1}{k}(\log q_{r+t})^k+O\left((\log q_{r+t})^{k-1}\right)\end{aligned}\tag{15}$$

Let $\omega(n)$ denote the number of distinct prime factors of n with $\omega(1)=0$. If $n>6$ then $n\ge 3^{\omega(n)}$ so that $\omega(n)<\log n$ for $n>6$.

By the prime number theorem, $q_r \sim r \log r$ as $r \to \infty$. Hence there exists a constant $M > 0$ such that

$$\begin{aligned} q_{r+t} &\leq M(r+t)\log(r+t) \\ &\leq M(t + \log n)\log(t + \log n) \\ &< M'(\log n).\log\log n, \end{aligned}$$

for large n. Hence for such n we have

$$\log q_{r+t} \leq \log\log n + O(\log\log\log n),$$

so that

$$\begin{aligned} (\log q_{r+t})^k &\leq (\log\log n + O(\log\log\log n))^k \\ &= (\log\log n)^k + O((\log\log n)^{k-1}(\log\log\log n)^k). \end{aligned}$$

Putting this into (15), we obtain lemma 5 from (14).

Lemma 6. *If* $1 < n = p_1^{\alpha_1} p_2^{\alpha_2} \ldots p_r^{\alpha_r}$ *is the canonical representation of n, then for large n, we have for any positive integer k,*

$$\sum_{i=1}^{r} S_k(p_i^{\alpha_i}) \leq \frac{1}{k}(\log\log n)^k + O((\log\log n)^{k-1}(\log\log\log n)^k).$$

Proof: We have

$$\begin{aligned} S_k(p^\alpha) &= \sum_{a=1}^{\alpha} \frac{(a\log p)^k}{p^a} = \frac{(\log p)^k}{p} + \sum_{a=2}^{\alpha} \frac{a^k(\log p)^k}{p^a} \\ &= (\log p)^k \left(\frac{1}{p} + \sum_{a=2}^{\alpha} \frac{a^k}{p^a} \right) = (\log p)^k \left(\frac{1}{p} + O\left(\frac{1}{p^2}\right)\right). \end{aligned}$$

Therefore we have

$$\begin{aligned} \sum_{i=1}^{r} S_k(p_i^{\alpha_i}) &\leq \sum_{i=1}^{r} \frac{(\log p_i)^k}{p_i} + O\left(\sum_{i=1}^{r} \frac{(\log p_i)^k}{p_i^2} \right) \\ &= \sum_{p|n} \frac{(\log p)^k}{p} + O\left(\sum_{p|n} \frac{(\log p)^k}{p^2} \right) \\ &= \sum_{p|n} \frac{(\log p)^k}{p} + O(1). \end{aligned}$$

Now lemma 6 follows from lemma 5.

3 Main Results

Theorem 1. *We have*

$$\overline{\lim_{n\to\infty}} \frac{1}{(\log\log n)^3} \sum_{d|n} \frac{(\log d)^2}{d} = \frac{3}{2}e^\gamma.$$

Proof: If $1 < n = p_1^{\alpha_1} p_2^{\alpha_2} \dots p_r^{\alpha_r}$, from lemma 2 and (6) ($k = 2$), we obtain

$$S_2(n) \le \left(\sum_{i=1}^{r} S_2(p_i^{\alpha_i}) + 2 \sum_{1 \le i < j \le r} S_1(p_i^{\alpha_i}) S_1(p_j^{\alpha_j}) \right) F(n). \tag{16}$$

For any positive integer α, we have

$$\sum_{a=1}^{\alpha} \frac{a}{p^a} \le \frac{1}{p} + \frac{6}{p^2}. \tag{17}$$

Also, by (5) ($k = 1$) and (17),

$$S_1(p^\alpha) = \sum_{a=1}^{\alpha} \frac{a \log p}{p^a} \le (\log p) \left(\frac{1}{p} + \frac{6}{p^2} \right). \tag{18}$$

Using (18), we obtain

$$\sum_{1 \le i < j \le r} S_1(p_i^{\alpha_i}) S_1(p_j^{\alpha_j}) \le \Sigma_1 + 6\Sigma_2 + 6\Sigma_3 + 36\Sigma_4, \tag{19}$$

where

$$\Sigma_1 = \sum_{1 \le i < j \le r} \frac{(\log p_i)(\log p_j)}{p_i p_j}, \tag{20}$$

$$\Sigma_2 = \sum_{1 \le i < j \le r} \frac{(\log p_i)(\log p_j)}{p_i p_j^2}, \tag{21}$$

$$\Sigma_3 = \sum_{1 \le i < j \le r} \frac{(\log p_i)(\log p_j)}{p_i^2 p_j}, \tag{22}$$

and

$$\Sigma_4 = \sum_{1 \le i < j \le r} \frac{(\log p_i)(\log p_j)}{p_i^2 p_j^2}. \tag{23}$$

We have

$$\begin{aligned} \left(\sum_{i=1}^{r} \frac{\log p_i}{p_i} \right)^2 &= \sum_{\substack{1 \le i \le r \\ 1 \le j \le r}} \frac{(\log p_i)(\log p_j)}{p_i p_j} \\ &= 2 \sum_{1 \le i < j \le r} \frac{(\log p_i)(\log p_j)}{p_i p_j} + \sum_{i=1}^{r} \left(\frac{\log p_i}{p_i} \right)^2. \end{aligned} \tag{24}$$

Therefore we have

$$\begin{aligned} 2 \sum_{1 \le i < j \le r} \frac{(\log p_i)(\log p_j)}{p_i p_j} &= \left(\sum_{i=1}^{r} \frac{\log p_i}{p_i} \right)^2 - \sum_{i=1}^{r} \left(\frac{\log p_i}{p_i} \right)^2 \\ &= \left(\sum_{p \mid n} \frac{\log p}{p} \right)^2 - \sum_{p \mid n} \frac{\log^2 p}{p^2} \end{aligned}$$

$$\begin{aligned} &\leq (\log\log n + O(\log\log\log n))^2 + O(1) \\ &= (\log\log n)^2 + O((\log\log n)(\log\log\log n)^2), \end{aligned} \tag{25}$$

by lemma 5 ($k = 1$).

From (21), we have

$$\begin{aligned} \Sigma_2 &= \sum_{2\le j\le r} \frac{\log p_j}{p_j^2} \sum_{1\le i<j} \frac{\log p_i}{p_i} \\ &= \sum_{2\le j\le r} \frac{\log p_j}{p_j^2} \sum_{p|n} \frac{\log p}{p} \\ &= O\left((\log\log n) \sum_{2\le j\le r} \frac{\log p_j}{p_j^2}\right) \\ &= O\left((\log\log n) \sum_{p|n} \frac{\log p}{p^2}\right) = O(\log\log n), \end{aligned} \tag{26}$$

where we used lemma 5 ($k = 1$).

Similarly we can show that

$$\Sigma_3 = O(\log\log n), \tag{27}$$

and

$$\Sigma_4 = O(1). \tag{28}$$

From (25)-(28) and (16), we obtain

$$S_2(n) \le \left(\frac{3}{2}(\log\log n)^2 + O\left((\log\log n).(\log\log\log n)^2\right)\right) F(n). \tag{29}$$

so that

$$\frac{S_2(n)}{(\log\log n)^3} \le \left(\frac{3}{2} + o(1)\right) \frac{F(n)}{\log\log n}. \tag{30}$$

Since we have (cf. [3], Theorem 3.23),

$$\overline{\lim_{n\to\infty}} \frac{F(n)}{\log\log n} = \overline{\lim_{n\to\infty}} \frac{\sigma(n)}{n\log\log n} = e^\gamma,$$

it follows from (30) that

$$\overline{\lim_{n\to\infty}} \frac{S_2(n)}{(\log\log n)^3} \le \frac{3}{2}e^\gamma. \tag{31}$$

We now prove the reverse inequality. Let

$$m_j = \prod_{p\le e^j} p^j, \tag{32}$$

for $j = 2, 3, 4, \ldots$. Let $p_1, p_2, \ldots, p_r$ denote all the distinct prime factors of m_j. From lemma 3 we have

$$\frac{S_2(m_j)}{F(m_j)} = \sum_{i=1}^{r} \frac{S_2(p_i^j)}{F(p_i^j)} + 2 \sum_{1\le i<k\le r} \frac{S_1(p_i^j)S_1(p_k^j)}{F(p_i^j)F(p_k^j)}. \tag{33}$$

For any positive integer α,

$$F(p^{\alpha}) = \sum_{d|p^{\alpha}} \frac{1}{d} = \sum_{i=o}^{\alpha} \frac{1}{p^i} \leq \sum_{i=o}^{\infty} \frac{1}{p^i} = \left(1 - \frac{1}{p}\right)^{-1},$$

so that

$$\frac{1}{F(p^{\alpha})} \geq \left(1 - \frac{1}{p}\right). \tag{34}$$

Using (34) in (33) we obtain,

$$\begin{aligned}
\frac{S_2(m_j)}{F(m_j)} &\geq \sum_{i=1}^{r} S_2(p_i^j)\left(1 - \frac{1}{p_i}\right) + 2 \sum_{1\leq i<k\leq r} S_1(p_i^j)S_1(p_k^j)\left(1 - \frac{1}{p_i}\right)\left(1 - \frac{1}{p_k}\right) \\
&\geq \sum_{i=1}^{r} \frac{\log^2 p_i}{p_i}\left(1 - \frac{1}{p_i}\right) + 2 \sum_{1\leq i<k\leq r} \frac{\log p_i}{p_i}\frac{\log p_k}{p_k}\left(1 - \frac{1}{p_i}\right)\left(1 - \frac{1}{p_k}\right) \\
&= \Sigma_5 + 2(\Sigma_1 - \Sigma_2 - \Sigma_3 + \Sigma_4),
\end{aligned} \tag{35}$$

say, where Σ_i for i=1 to 4 are as given in (20)–(23) with the running symbol j replaced by k. In the above we used that $S_2(p^{\alpha}) \geq \frac{\log^2 p}{p}$ and $S_1(p^{\alpha}) \geq \frac{\log p}{p}$.

By lemma 4 ($k = 2$), we have

$$\begin{aligned}
\Sigma_5 &= \sum_{p|m_j} \frac{\log^2 p}{p} - \sum_{p|m_j} \frac{\log^2 p}{p^2} = \sum_{p\leq e^j} \frac{\log^2 p}{p} + O(1) \\
&= \frac{1}{2}(\log e^j)^2 + O(j) + O(1). \\
&= \frac{1}{2}j^2 + O(j).
\end{aligned} \tag{36}$$

From (24) we have

$$\begin{aligned}
2\Sigma_1 &= \left(\sum_{p|m_j} \frac{\log p}{p}\right)^2 - \sum_{p|m_j} \frac{\log^2 p}{p^2} \\
&= \left(\sum_{p\leq e^j} \frac{\log p}{p}\right)^2 + O(1) \\
&= j^2 + O(j).
\end{aligned} \tag{37}$$

From (26)-(28) we obtain,

$$\Sigma_2 = O(\log\log m_j), \tag{38}$$

$$\Sigma_3 = O(\log\log m_j), \tag{39}$$

and

$$\Sigma_4 = O(1). \tag{40}$$

From (40)-(36) and (35), we obtain

$$S_2(m_j) \geq \left(\frac{3}{2}j^2 + O(j) + O(\log\log m_j)\right)F(m_j),$$

so that

$$\frac{S_2(m_j)}{(\log\log m_j)^3} \geq \left(\frac{3}{2}\frac{j^2}{(\log\log m_j)^2} + O\left(\frac{j}{(\log\log m_j)^2}\right) + O\left(\frac{1}{(\log\log m_j)}\right)\right)\frac{F(m_j)}{\log\log m_j}. \tag{41}$$

Using the Prime number theorem in the form $\theta(x) = \sum_{p\leq x}\log p \sim x$ as $x \to \infty$, we obtain from (32),

$$\log m_j = j\theta(e^j) \sim je^j \ (j \to \infty),$$

so that

$$\log\log m_j \sim j + \log j \sim j \ (j \to \infty). \tag{42}$$

Also, it is known (cf. [3], Theorem 3.23) that

$$\lim_{j\to\infty}\frac{F(m_j)}{\log\log m_j} = \lim_{j\to\infty}\frac{\sigma(m_j)}{m_j\log\log m_j} = e^\gamma. \tag{43}$$

From (42) and (43), it follows that the right hand side of (41) tends to $\frac{3}{2}e^\gamma$ as $j \to \infty$. Hence we obtain

$$\overline{\lim_{n\to\infty}}\frac{S_2(n)}{(\log\log n)^3} \geq \frac{3}{2}e^\gamma.$$

This completes the proof of Theorem 1.

Theorem 2. *We have*

$$\overline{\lim_{n\to\infty}}\frac{1}{(\log\log n)^4}\sum_{d|n}\frac{\log^3 d}{d} = \frac{17}{6}e^\gamma.$$

Proof: Let $1 < n = p_1^{\alpha_1}p_2^{\alpha_2}\dots p_r^{\alpha_r}$ be the canonical representation. Then by lemma 3 we have

$$B_3(n) = \Sigma_6 + 3\Sigma_7 + 6\Sigma_8, \tag{44}$$

where

$$\begin{aligned}\sum_6 &= \sum_{i=1}^r B_3(p_i^{\alpha_i}) = \sum_{i=1}^r \frac{S_3(p_i^{\alpha_i})}{F(p_i^{\alpha_i})}\\ &\leq \sum_{i=1}^r S_3(p_i^{\alpha_i})\\ &\leq \frac{1}{3}(\log\log n)^3 + O((\log\log n)^2(\log\log\log n)^3),\end{aligned} \tag{45}$$

by lemma 6 ($k = 2$).

We have again by lemma 6,

$$\begin{aligned}\Sigma_7 &= \sum_{\substack{1\leq i\leq r\\1\leq j\leq r\\i\neq j}} B_2(p_i^{\alpha_i})B_1(p_j^{\alpha_j})\\ &\leq \left(\sum_{i=1}^r B_2(p_i^{\alpha_i})\right)\left(\sum_{j=1}^r B_1(p_j^{\alpha_j})\right)\end{aligned}$$

$$
\begin{aligned}
&\le \left(\sum_{i=1}^{r} S_2(p_i^{\alpha_i})\right)\left(\sum_{j=1}^{r}) S_1(p_j^{\alpha_j})\right) \\
&\le \left(\frac{1}{2}(\log\log n)^2 + O((\log\log n)(\log\log\log n)^2\right)(\log\log n + O(\log\log\log n)) \\
&= \frac{1}{2}(\log\log n)^3 + O\left((\log\log n)^2(\log\log\log n)^3\right). \qquad (46)
\end{aligned}
$$

We have

$$
\begin{aligned}
\Sigma_8 &= \sum_{1\le i<j<k\le r} B_1(p_i^{\alpha_i})B_1(p_j^{\alpha_j})B_1(p_k^{\alpha_k}) \\
&\le \sum_{1\le i<j<k\le r} \left(\frac{\log p_i}{p_i} + 6\frac{\log p_i}{p_i^2}\right)\left(\frac{\log p_j}{p_j} + 6\frac{\log p_j}{p_j^2}\right)\left(\frac{\log p_k}{p_k} + 6\frac{\log p_k}{p_k^2}\right) \\
&= \sum_{1\le i<j<k\le r} \frac{(\log p_i)(\log p_j)\log p_k}{p_i p_j p_k} + O((\log\log n)^2). \qquad (47)
\end{aligned}
$$

We have

$$
\begin{aligned}
\left(\sum_{p|n} \frac{\log p}{p}\right)^3 &= \sum_{\substack{1\le i\le r\\ 1\le j\le r\\ i\le k\le r}} \frac{(\log p_i)(\log p_j)(\log p_k)}{p_i p_j p_k} \\
&= 6 \sum_{1\le i<j<k\le r} \frac{(\log p_i)(\log p_j)(\log p_k)}{p_i p_j p_k} \\
&\quad + 3\left(\sum_{i=1}^{r} \frac{\log^2 p_i}{p_i^2}\right)\left(\sum_{k=1}^{r} \frac{\log p_k}{p_k}\right) - 2\sum_{i=1}^{r} \frac{\log^3 p_i}{p_i^3}. \qquad (47')
\end{aligned}
$$

Therefore by lemma 5,

$$
\begin{aligned}
6 \sum_{1\le i<j<k\le r} \frac{(\log p_i)(\log p_j)(\log p_k)}{p_i p_j p_k} &= \left(\sum_{p|n} \frac{\log p}{p}\right)^3 \\
&\quad -3\left(\sum_{i=1}^{r} \frac{\log^2 p_i}{p_i^2}\right)\left(\sum_{k=1}^{r} \frac{\log p_k}{p_k}\right) + 2\sum_{p|n} \frac{\log^3 p}{p^3} \\
&\le [\log\log n + O(\log\log\log n)]^3 + O(\log\log n) + O(1) \\
&= (\log\log n)^3 + O((\log\log n)^2(\log\log\log n)^3). \qquad (48)
\end{aligned}
$$

Hence from (47) and (48),

$$
6\,\Sigma_8 \le (\log\log n)^3 + O((\log\log n)^2(\log\log\log n)^3). \qquad (49)
$$

Now from (45),(46), (49)and (44) we obtain

$$
B_3(n) \le \frac{17}{6}(\log\log n)^3 + O((\log\log n)^2(\log\log\log n)^3),
$$

so that

$$S_3(n) \le \left(\frac{17}{6}(\log\log n)^3 + O((\log\log n)^2(\log\log\log n)^3)\right) F(n)$$

and hence

$$\overline{\lim_{n\to\infty}} \frac{S_3(n)}{(\log\log n)^4} \le \frac{17}{6} e^{\gamma}.$$

Let m_j be given as in (32) and $p_1, p_2, \ldots, p_r$ be the distinct prime factors of m_j. By lemmas 3 and (24) we have

$$\begin{aligned} B_3(m_j) &= \sum_{i=1}^{r} B_3(p_i^j) + 3 \sum_{\substack{1\le i\le r\\ 1\le k\le r\\ i\ne k}} B_2(p_i^j)B_1(p_k^j) + 6 \sum_{1\le i<\ell<k\le r} B_1(p_i^j)B_1(p_\ell^j)B_1(p_k^j). \\ &\ge \sum_{i=1}^{r} \frac{\log^3 p_i}{p_i}\left(1-\frac{1}{p_i}\right) + 3\sum_{\substack{1\le i\le r\\ 1\le k\le r}} B_2(p_i^j)B_1(p_k^j) - 3\sum_{1\le i\le r} B_2(p_i^j)B_1(p_i^j) \\ &\quad +6 \sum_{1\le i<\ell<k\le r} \frac{\log p_i}{p_i}.\frac{\log p_\ell}{p_\ell}.\frac{\log p_k}{p_k}\left(1-\frac{1}{p_i}\right)\left(1-\frac{1}{p_\ell}\right)\left(1-\frac{1}{p_k}\right) \end{aligned}$$

Continuing, we have

$$\begin{aligned} B_3(m_j) &\ge \sum_{p\le e^j} \frac{\log^3 p}{p} + O(1) + 3\left(\sum_{1\le i\le r} \frac{\log^2 p_i}{p_i}\left(1-\frac{1}{p_i}\right)\right) \\ &\quad \left(\sum_{1\le k\le r} \frac{\log p_k}{p_k}\left(1-\frac{1}{p_k}\right)\right) - 3\sum_{1\le i\le r} B_2(p_i^j)B_1(p_i^j) \\ &\quad + 6 \sum_{1\le i<\ell<k\le r} \frac{\log p_i}{p_i}.\frac{\log p_\ell}{p_\ell}.\frac{\log p_k}{p_k} + O((\log\log m_j)^2). \end{aligned} \tag{50}$$

We have

$$\begin{aligned} \sum_{1\le i\le r} B_2(p_i^j)B_1(p_i^j) &\le \sum_{i=1}^{r} (\log^3 p_i)\left(\frac{1}{p_i} + O\left(\frac{1}{p_i^2}\right)\right)^2 \\ &= O\left(\sum_{i=1}^{r} \frac{\log^3 p_i}{p_i^2}\right) = O(1). \end{aligned} \tag{51}$$

We have from (47′), lemmas 5 and 4 ($k = 1$),

$$\begin{aligned} 6 \sum_{1\le i<\ell<k\le r} \frac{\log p_i}{p_i}.\frac{\log p_\ell}{p_\ell}.\frac{\log p_k}{p_k} &= \left(\sum_{p\mid m_j} \frac{\log p}{p}\right)^3 - 3\left(\sum_{p\mid m_j} \frac{\log^2 p}{p^2}\right) \\ &\quad \left(\sum_{p\mid m_j} \frac{\log p}{p}\right) + 2\sum_{p\mid m_j} \frac{\log^3 p}{p^3} \\ &= \left(\sum_{p\le e^j} \frac{\log p}{p}\right)^3 + O(\log\log m_j) \end{aligned}$$

$$
\begin{aligned}
&= (j+O(1))^3 + O(\log\log m_j) \\
&= j^3 + O(j^2) + O(\log\log m_j). \qquad (52)
\end{aligned}
$$

Using (51) and (52) in (50), we obtain by lemma 4,

$$
\begin{aligned}
B_3(m_j) &= \frac{S_3(m_j)}{F(m_j)} \geq \sum_{p\leq e^j} \frac{\log^3 p}{p} + 3\left(\sum_{p\leq e^j} \frac{\log^2 p}{p} + O(1)\right)\left(\sum_{p\leq e^j} \frac{\log p}{p} + O(1)\right) \\
&\quad + O(1) + j^3 + O(j^2) + O(\log\log m_j) + O((\log\log m_j)^2) \\
&= \frac{17}{6} j^3 + O(j^2) + O((\log\log m_j)^2),
\end{aligned}
$$

so that

$$
\begin{aligned}
\frac{S_3(m_j)}{(\log\log m_j)^4} &\geq \left(\frac{17}{6}\frac{j^3}{(\log\log m_j)^3} + O\left(\frac{j^2}{(\log\log m_j)^3}\right)\right. \\
&\quad \left. + O\left(\frac{1}{\log\log m_j}\right)\right)\frac{F(m_j)}{\log\log m_j} \\
&\to \frac{17}{6} e^{\gamma},
\end{aligned}
$$

by (42) and (43). This completes the proof of Theorem 2.

4 Unitary analogues

For each positive integer k, let

$$
S_k^*(n) = \sum_{d\|n} \frac{(\log d)^k}{d}
$$

and

$$
B_k^*(n) = \frac{S_k^*(n)}{F^*(n)},
$$

where

$$
F^*(n) = \sum_{d\|n} \frac{1}{d} = \frac{\sigma^*(n)}{n},
$$

$\sigma^*(n)$ being the sum of the unitary divisors of n ; in the above $d\|n$ means that d is a divisor of n and is relatively prime to n/d i.e., d is a unitary divisor of n(cf. [1]). It is not difficult to show that Lemmas 2 and 3 hold when B_1,B_2 and B_3 are replaced by B_1^*,B_2^* and B_3^* respectively. If we use that $S_k^*(n) \leq S_k(n)$,

$$
\overline{\lim_{n\to\infty}} \frac{\sigma^*(n)}{n\log\log n} = \frac{6}{\pi^2} e^{\gamma}, \qquad (53)
$$

and

$$
\lim_{j\to\infty} \frac{\sigma^*(n_j)}{n_j\log\log n_j} = \frac{6}{\pi^2} e^{\gamma}, \qquad (54)
$$

in the proofs of Theorems 1 and 2, where

$$n_j = \prod_{p \le j} p,$$

we obtain the following:

$$\overline{\lim_{n\to\infty}} \frac{1}{(\log\log n)^3} \sum_{d\|n} \frac{(\log d)^2}{d} = \frac{9}{\pi^2} e^{\gamma},$$

and

$$\overline{\lim_{n\to\infty}} \frac{1}{(\log\log n)^4} \sum_{d\|n} \frac{\log^3 d}{d} = \frac{17}{\pi^2} e^{\gamma}.$$

For the proofs of (53) and (54) we refer to S.Wigert [6]: We may note that the result

$$\overline{\lim_{n\to\infty}} \frac{1}{(\log\log n)^2} \sum_{d\|n} \frac{\log d}{d} = \frac{6}{\pi^2} e^{\gamma},$$

has been established in [5].

Addendum: Taking into account the suggestions expressed by a referee to remain anonymous, we would like to add the following: Let $s > 0$ and

$$f(s) = \sum_{d|n} \frac{1}{d^s} = \prod_{p|n} \left(1 + \frac{1}{p^s} + \frac{1}{p^{2s}} \cdots + \frac{1}{p^{\alpha s}}\right). \tag{55}$$

Then $S_k(n) = (-1)^k f^{(k)}(1)$ where $S_k(n)$ is as given in (5) and $f^{(k)}$ is the $k - th$ derivative of f. Logarithmic differentiation of (55) gives

$$\begin{aligned} \frac{f'(s)}{f(s)} &= \sum_{p^{\alpha}\|n} \frac{(\alpha+1)\log p}{p^{(\alpha+1)s} - 1} - \sum_{p|n} \frac{\log p}{p^s - 1} \\ &= \left(-\sum_{p|n} \frac{\log p}{p^s} + M(s)\right) = g(s), \end{aligned} \tag{56}$$

say, where

$$M(s) = \sum_{p^{\alpha}\|n} \frac{(\alpha+1)\log p}{p^{(\alpha+1)s} - 1} - \sum_{p|n} \frac{\log p}{p^s(p^s - 1)}. \tag{57}$$

From (56), we have $f'(s) = f(s)g(s)$; differentiating this k times and putting $s = 1$, we obtain

$$f^{(k+1)}(1) = \sum_{r=0}^{k} \binom{k}{r} g^{(r)}(1) f^{(k-r)}(1). \tag{58}$$

Now, we assume as induction hypothesis that for a given $\epsilon > 0$ and for all $0 \le \ell \le k$,

$$S_{\ell}(n) = (-1)^{\ell} f^{(\ell)}(1) \le e^{\gamma} C_{\ell}(1+\epsilon)(\log\log n)^{n+1},$$

for all $n \ge N(\ell)$ and

$$S_{\ell}(n) = (-1)^{\ell} f^{(\ell)}(1) \le e^{\gamma} C_{\ell}(1-\epsilon)(\log\log n)^{n+1},$$

when $n = \prod_{p \le e^j} p^j$, (as $j \to \infty$).Since

$$\sum_{p|n} \frac{(\log p)^{r+1}}{r+1} \le \frac{(\log\log n)^{r+1}}{r+1}(1+\varepsilon),$$

for large n (Lemma 5) and

$$M^{(r)}(1) = O((\log\log n)^r),$$

we obtain from (58) and the induction hypothesis

$$0 \le \frac{S_{k+1}(n)}{(\log\log n)^{k+2}} \le e^{\gamma}\left(\sum_{r=0}^{k}\binom{k}{r}(r+1)^{-1}C_{k-r}\right)(1+\varepsilon),$$

for all $n \ge N(k)$ and

$$\frac{S_{k+1}(n)}{(\log\log n)^{k+2}} \ge e^{\gamma}\left(\sum_{r=0}^{k}\binom{k}{r}(r+1)^{-1}C_{k-r}\right)(1-\varepsilon),$$

if $n = \prod_{p \le e^j} p^j (j \to \infty)$. This proves that

$$\overline{\lim_{n\to\infty}} \frac{S_{k+1}(n)}{(\log\log n)^{k+2}} = C_{k+1}e^{\gamma},$$

where

$$C_{k+1} = \sum_{r=0}^{k}\binom{k}{r}(r+1)^{-1}C_{k-r}.$$

(Since $C_0 = C_1 = 1$, we get $C_2 = \frac{3}{2}$ and $C_3 = \frac{17}{6}$.) The conjecture (2) is now settled. In a similar way we can prove that

$$\overline{\lim_{n\to\infty}}(\log\log n)^{-(k+1)} \sum_{d\|n} \frac{(\log d)^k}{d} = \frac{6}{\pi^2}C_k e^{\gamma}.$$

References

[1] E. Cohen, Arithmetical functions associated with the unitary divisors of an integer, *Math.Z.* 74 (1960), 66-80.

[2] P. Erdös and S.K. Zaremba, The arithmetic function $\sum_{d|n} \frac{\log d}{d}$, *Demonstration Math.* 6 (1973), 575-579.

[3] G.H. Hardy and E.M. Wright, *An introduction to the theory of numbers*, Fifth edition, Oxford Univ. 1987.

[4] Expansion of $\sum_{p \le x} \frac{(\log p)^k}{p}$, *Bull.Iranian Math Soc.* 14 (1987), 31-39.

[5] V. Sitaramaiah and M.V. Subbarao, The maximal order of certain arithmetical functions, *Indian J.Pure Appl. Math.*, 24 (1993), 347-355.

[6] S. Wigert, Note sur Deux function arithmetiques, *Prace Matematyczno-Fizyczne*, 38 (1931), 23-29.

V. Sitaramaiah
Department of Mathematics,
Pondicherry Engineering College,
Pondicherry, 605 014,
India.

e-mail: ramaiahpec@yahoo.co.in

M.V. Subbarao
Department of Mathematical Sciences,
University of Alberta, Edmonton,
Alberta,
Canada T6G2G1.

e-mail: msubbara@ualberta.ca

The Riemann Zeta Function and Related Themes – 2006, pp. 155–179

On Ramachandra's contributions to transcendental number theory

Michel Waldschmidt

Dedicated to Professor K. Ramachandra on his 70th birthday

The title of this lecture refers to Ramachandra's paper in Acta Arithmetica [36], which will be our central subject: in section 1 we state his Main Theorem, in section 2 we apply it to algebraically additive functions. Next we give new consequences of Ramachandra's results to density problems; for instance we discuss the following question: *let E be an elliptic curve which is defined over the field of algebraic numbers, and let Γ be a finitely generated subgroup of algebraic points on E; is Γ dense in $E(\mathbb{C})$ for the complex topology?* The other contributions of Ramachandra to transcendental number theory are dealt with more concisely in section 4. Finally we propose a few open problems.

The author wishes to convey his best thanks to the organizer of the Madras Conference of July 1993 in honor of Professor Ramachandra's 60th birthday, R. Balasubramanian, for his invitation to participate, which provided him the opportunity to write this paper. Next he is grateful to the organizer of the Bangalore Conference of December 2003 in honor of Professor Ramachandra's 70th birthday, K. Srinivas, for his invitation to participate, which provided him the opportunity to publish this paper. He is also glad to express his deep gratitude to Professor K. Ramachandra for the inspiring role of his work and for his invitation to the Tata Institute as early as 1976.

1 Ramachandra's Main Theorem

Hilbert's seventh problem on the transcendence of α^β (for algebraic α and β) was solved in 1934 by Gel'fond and Schneider, using two different approaches: while Gel'fond's solution [14] involved the differential equation $(d/dz)e^z = e^z$ of the exponential function, Schneider's proof [45] rested on the addition formula $e^{z_1+z_2} = e^{z_1}e^{z_2}$. Later, both methods were developed and applied to other functions, notably the elliptic functions. In particular Schneider in [46] proved an elliptic analog of the theorem on the transcendence of α^β, using the differential equation which is satisfied by a Weierstrass elliptic function: $\wp'^2 = 4\wp^3 - g_2\wp - g_3$. Sometimes, one refers to *Schneider's method* when no derivative is needed, and to *Gel'fond's method* when differential equations are there; but, as pointed out by A. Baker, this terminology is somewhat deficient, since for instance Schneider's early results on elliptic functions [46] involve derivatives, and furthermore the first result on functions of several variables (which yields the transcendence of the values of the Beta function at rational points) has been proved by Schneider in [47] using a variant of Gel'fond's method!

The first general criterion dealing with analytic or meromorphic functions of one variable and containing the solution to Hilbert's seventh problem appears in [48]; in fact one can deduce the transcendence of α^β from this criterion by both methods, either by using the two functions z and α^z (Schneider's method), or else e^z and $e^{\beta z}$ (Gel'fond's method). This criterion is somewhat complicated, and Schneider made successful attempts to simplify it [49]; however these last results deal only with Gel'fond's method, i.e. derivatives are needed. A further simplification for functions satisfying differential equations was provided by Lang later ([21] and [22]); the so-called *Schneider-Lang criterion* was used by Bertrand and Masser to derive Baker's Theorem on linear independence of logarithms, as well as its elliptic analog [5]; also it was extended to functions of several variables by Bombieri, solving a Conjecture of Nagata [6].

Thus the situation for functions satisfying differential equations (Gel'fond's method) is rather satisfactory; but it is not the same for Schneider's method. The difficulty of providing simple criteria without assuming differential equations is illustrated by examples due to Weierstrass, Stäckel and others (see [29]). The work of Ramachandra which we consider here deals with this question. Simple criteria are known, the first one being Pólya's Theorem: *there is no entire function which is not a polynomial, which maps the natural integers into* $\mathbb{Z}$, *and which has a growth order less than* 2^z (see [15] Chap. III, §2, for related results, [60] and [63] for surveys, and [59] for a proof which is inspired by Ramachandra's work).

The first part of [36] contains an introduction, the statement of some results and of the Main Theorem, and the proof of it. The second part is devoted to corollaries of the Main Theorem. We reproduce here the Main Theorem.

We denote by $\overline{\mathbb{Q}}$ the field of complex algebraic numbers (algebraic closure of $\mathbb{Q}$ in $\mathbb{C}$). The *size* of an algebraic number α is defined by $\operatorname{size}\alpha = \operatorname{den}\alpha + \overline{|\alpha|}$, where $\operatorname{den}\alpha$ is the denominator of α (the least natural integer d such that $d\alpha$ is an algebraic integer) and $\overline{|\alpha|}$ is the *house* of α (maximum of the absolute values of the complex conjugates of α). We also need the following definition: an entire function f in $\mathbb{C}$ is *of order* $\leq \varrho$ if there exists $C > 0$ such that for $R \geq 1$

$$\log \sup_{|z|=R} |f(z)| \leq CR^{\varrho}.$$

(1) Let $d \geq 2$ be a natural number and ϱ a positive real number; for $1 \leq i \leq d$, let g_i and h_i be two entire functions without common zeros, of order $\leq \varrho$, and let $M^{(i)}(R)$ denote the quantity

$$M^{(i)}(R) = \left(1 + \max_{|z|=R} |h_i(z)|\right)\left(1 + \max_{|z|=R} |g_i(z)|\right).$$

Assume further that the d meromorphic functions $f_i = h_i/g_i$, $(1 \leq i \leq d)$ are algebraically independent over $\mathbb{C}$.

(2) Let $(\zeta_\mu)_{\mu\geq 1}$ be an infinite sequence of distinct complex numbers, and $(n_\mu)_{\mu\geq 1}$ be a non-decreasing sequence of natural numbers with $\lim_{\mu\to\infty} n_\mu = \infty$. For $Q \geq 1$, define

$$N(Q) = \operatorname{Card}\{\mu;\ \mu \geq 1,\ n_\mu \leq Q\} \qquad \text{and} \qquad D(Q) = \max_{n_\mu \leq Q} |\zeta_\mu|,$$

and assume

$$\liminf_{Q\to\infty} \frac{\log N(Q)}{\log D(Q)} > \varrho.$$

(3) Let $(\mu_r)_{r\geq 1}$ be a sequence of integers such that the number

$$N_1(Q) = \operatorname{Card}\{\mu_r;\ r \geq 1,\ n_{\mu_r} \leq Q\}$$

tends to infinity as Q tends to infinity. Suppose that whenever a polynomial in $f_1, \ldots, f_d$ vanishes at all points ζ_{μ_r} with $n_{\mu_r} \leq Q$, then it vanishes also at all points ζ_μ with $n_\mu \leq Q$.

(4) Suppose that the numbers $f_i(\zeta_\mu)$ (for $1 \le i \le d$ and $\mu \ge 1$) are all algebraic numbers; denote by $\partial(Q)$ the degree of the field obtained by adjoining the algebraic numbers

$$f_i(\zeta_\mu), \qquad (1 \le i \le d, \quad 1 \le \mu \le Q)$$

to the field of rational numbers, and set

$$M_1^{(i)}(Q) = 1 + \max_{n_\mu \le Q}\{\mathrm{size}(f_i(\zeta_\mu))\}, \qquad (1 \le i \le d).$$

(5) Finally set

$$M_2^{(i)}(Q) = 1 + \max_{n_\mu \le Q} \frac{1}{|g_i(\zeta_\mu)|}, \qquad (1 \le i \le d).$$

Main Theorem of [Ramachandra 1968]. *Let q be a sufficiently large natural number, and $L_1, \dots, L_d$ natural numbers related to q asymptotically by*

$$L_1 \cdots L_d \sim \partial(q)(\partial(q) + 1)N_1(q).$$

Suppose that the hypotheses (1) — (4) above are satisfied. Then there exists a natural number Q, greater than q, such that for every positive quantity R, there holds

$$1 \le \left(\frac{8D(Q)}{R}\right)^{N(Q-1)} \prod_{i=1}^{s}\left((M_1^{(i)}(Q))^{8\partial(Q)} M_2^{(i)}(Q) M^{(i)}(R)\right)^{L_i}.$$

Remark: It would be interesting to write down a proof of this result by means of Laurent's interpolation determinants (see [24, 25, 26, 27] and [64]): instead of using Dirichlet's box principle (lemma of Thue-Siegel) for constructing an auxiliary function, one considers the matrix of the related system of equations, and one estimates (from below using Liouville's inequality, from above thanks to Schwarz's lemma) a non-vanishing determinant. It is to be expected that such an argument will produce a slightly different explicit inequality, but it is unlikely that these differences will have any effect on the corollaries.

Further works in this direction have been developed; see in particular [57, 30, 59, 16, 17, 34].

The paper [7] can be considered as the first extension of Ramachandra's Theorem to higher dimension; more recent results connected with functions of several variables are given in [62].

2 Pseudo-algebraic points of algebraically additive functions

a) Statement of Ramachandra's upper bound

When f is a meromorphic function in the complex plane and y is a complex number, we shall say that *y is a pseudo-algebraic point of f* if either y is a pole of f or else $f(y)$ is an algebraic number (this is the definition in [36] p. 84).

Let $f_1, \dots, f_d$ be meromorphic functions; we define $\delta(f_1, \dots, f_d)$ (which is either a non-negative integer or else ∞) as the dimension of the space of pseudo-algebraic point of f_i.

This notation is convenient to state a few classical transcendence results: the Theorem of Hermite-Lindemann is $\delta(z, e^z) = 0$, the Theorem of Gel'fond-Schneider can be stated either as

– *(Gel'fond's method): for any irrational algebraic number* β, $\delta(e^z, e^{\beta z}) = 0$;

– *(Schneider's method): for any non-zero complex number* t, $\delta(z, e^{tz}) \leq 1$.

Schneider's results on elliptic functions (see for instance [49] Chapitre 2 §3 Théorèmes 15, 16 et 18) can also be stated as follows:

if $\wp$ *and* $\wp^*$ *are Weierstrass elliptic functions with algebraic invariants* g_2, g_3, g_2^*, g_3^**, if* β *and* γ *are non-zero algebraic numbers such that the two functions* $\wp(z)$ *and* $\wp^*(\gamma z)$ *are algebraically independent, if* ζ *is the Weierstrass zeta function associated to* $\wp$*, and if* a, b *are algebraic numbers with* $(a, b) \neq (0, 0)$*, then*

$$\delta(e^{\beta z}, \wp(z)) = 0, \qquad \delta(\wp(z), az + b\zeta(z)) = 0$$

and

$$\delta(\wp(z), \wp^*(\gamma z)) = 0. \tag{1}$$

These results of Schneider on elliptic functions depend heavily on the fact that these functions satisfy differential equations with algebraic coefficients. The main point in Ramachandra's work is that similar results are achieved for functions which do not satisfy differential equations with algebraic coefficients. In place of derivatives, the addition theorem which is satisfied by these functions plays a crucial role. According to [36] p. 85, a meromorphic function f is said to possess an *algebraic addition theorem* if there exists a non-zero polynomial $P \in \mathbb{C}[T_1, T_2, T_3]$ such that the meromorphic function of three variables $P(f(z_1 + z_2), f(z_1), f(z_2))$ is the zero function. Further, if there is such a polynomial P with algebraic coefficients, then f will be called *algebraically additive*.

If $f_1, \ldots, f_d$ are algebraically additive functions, then the set of common pseudo-algebraic points of $f_1, \ldots, f_d$ can be shown to be a $\mathbb{Q}$-vector space, and $\delta(f_1, \ldots, f_d)$ is nothing else than the dimension of this vector space. The fundamental result in part II of [36] (Theorem 1 p. 74) is the following upper bound for this dimension:

Ramachandra's δ–Theorem. – *Let* $f_1, \ldots, f_d$*, with* $d \geq 2$*, be algebraically independent meromorphic functions; assume that for* $1 \leq i \leq d$*, the function* f_i *is algebraically additive and is of order* $\leq \varrho_i$*. Define*

$$\kappa = \begin{cases} 1 & \text{if } f_1, \ldots, f_d \text{ have a common non-zero period,} \\ 0 & \text{otherwise.} \end{cases}$$

Then

$$\delta(f_1, \ldots, f_d) \leq \frac{\varrho_1 + \cdots + \varrho_d - \kappa}{d - 1}.$$

Lang's criterion for Schneider's method in [22] (Chapter 2, Theorem 2) is the following special case:

$$\text{For } d = 2 \text{ the inequality } \delta(f_1, f_2) \leq 2\varrho^* \text{ holds with } \varrho^* = \max\{\varrho_1, \varrho_2\}. \tag{2}$$

We quote from [36] p. 87: "*It may be possible to improve the bound for the dimension given by Theorem 2 probably to 1 in all cases; but even a slight improvement such as* $\leq \varrho^* + (\varrho_* - \kappa)/d$ *appears to be very difficult*". (We have substituted κ and d to θ and s respectively to cope with our own notations). The number ϱ^* stands for $\max\{\varrho_1, \ldots, \varrho_d\}$, while ϱ_* stands for $\min\{\varrho_1, \ldots, \varrho_d\}$.

It is quite remarquable that no substantial improvement of Ramachandra's δ–Theorem has been obtained after more than a quarter of a century!

We now describe one situation where the assumption of Ramachandra's δ – Theorem are satisfied, and nevertheless the estimate $\delta(f_1,\ldots,f_d)\le 1$ does not hold. Take a Weierstrass elliptic function $\wp$ with algebraic invariants g_2 and g_3; let t be a non-zero complex number; consider the two functions $f_1(z)=z$ and $f_2(z)=\wp(tz)$. From Ramachandra's δ–theorem follows $\delta(z,\wp(tz))\le 2$. To start with, assume equality holds: let α and β be two $\mathbb{Q}$-linearly independent common pseudo-algebraic points of f_1 and f_2; then $u=t\alpha$ and $v=t\beta$ are $\mathbb{Q}$-linearly independent common pseudo-algebraic points of $\wp$, and the quotient $\gamma=u/v=\alpha/\beta$ is algebraic irrational; hence $\delta(\wp(z),\wp(\gamma z))\ge 1$; from Schneider's above mentioned theorem (1) with $\wp^*=\wp$, we deduce that the two functions $\wp(z)$ and $\wp(\gamma z)$ are algebraically dependent; now γ is irrational, hence we are in the so-called "CM case": when (ω_1,ω_2) is a fundamental pair of periods, $\tau=\omega_2/\omega_1$ is an imaginary quadratic number and the associated elliptic curve has a non trivial ring of endomorphisms.

Conversely, if $\wp$ has complex multiplications, for any $t\in\mathbb{C}^\times$ the set of common pseudo-algebraic points of the two functions z and $\wp(tz)$ is a $\mathbb{Q}(\tau)$-vector space, and therefore $\delta(z,\wp(tz))$ is even; one example where this vector space has positive dimension is when t is a rational multiple of a period of $\wp$; for instance

$$\delta(z,\wp(\omega_1 z))=2.$$

We repair Ramachandra's Conjecture as follows:

Ramachandra's δ–Conjecture. – *For any non-zero complex number t and any Weierstrass elliptic function $\wp$ (resp. $\wp^*$) with algebraic invariants g_2,g_3 (resp. g_2^*,g_3^*), assuming the two functions $\wp(tz)$ and $\wp^*(z)$ are algebraically independent,*

$$\delta(e^z,e^{tz})\le 1,\qquad \delta(e^z,\wp(tz))\le 1,\qquad \delta(\wp(tz),\wp^*(z))\le 1.$$

The first inequality is equivalent to the so-called *four exponentials Conjecture*, which was apparently known to Siegel (see [1]), which was also considered by A. Selberg in the early 40's (personal communication, Hong-Kong, July 1993) and later was proposed by Lang in [22]; an equivalent question is the first of Schneider's eight problems in [49]:

Four exponentials Conjecture. – *Let x_1,x_2 be two $\mathbb{Q}$-linearly independent complex numbers, and y_1,y_2 be also two $\mathbb{Q}$-linearly independent complex numbers; then one at least of the four numbers*

$$e^{x_1y_1},\quad e^{x_1y_2},\quad e^{x_2y_1},\quad e^{x_2y_2}$$

is transcendental.

A partial result can be proved (which is called *the five exponentials Theorem* in [62]; the fifth number is e^{x_1/x_2}). A result which is stronger than both the six and the five exponentials Theorems, but which does not include the four exponentials Conjecture, is due to D. Roy [43]). Denote by $\tilde{\mathcal{L}}$ the set of linear combinations of logarithms of algebraic numbers, which is the $\overline{\mathbb{Q}}$-vector space spanned by $\{1\}\cup\{\ell\in\mathbb{C};\ e^\ell\in\overline{\mathbb{Q}}^\times\}$.

Theorem (D. Roy). – *If* x_1, x_2, x_3 *are* $\overline{\mathbb{Q}}$*-linearly independent complex numbers and* y_1, y_2 *are* $\overline{\mathbb{Q}}$*-linearly independent complex numbers, then one at least of the six numbers* $x_i y_j$ *is not in* $\tilde{\mathcal{L}}$.

b) Sketch of proof of Ramachandra's δ–Theorem as a consequence of Ramachandra's Main Theorem

The deduction of the δ–Theorem from the Main Theorem is by no means trivial; Ramachandra had to use Weyl's criterion of equidistribution in order to check the hypotheses concerning the poles of the elliptic functions. Another approach is due to Serre (see lemma 3.4 p. 46 of Waldschmidt, 1973]). Here, in this sketch of proof, we hardly quote problems arising from the poles.

Inside the group of common pseudo-algebraic points of $f_1, \dots, f_d$, select a finitely generated subgroup $Y = \mathbb{Z}y_1 + \cdots + \mathbb{Z}y_\ell$ of rank ℓ. In case $\kappa = 1$, choose for y_1 a common period to $f_1, \dots, f_d$. Let $h_1, h_2, \dots$ be a numbering of $\mathbb{Z}^\ell$, with $h_\mu = (h_{1\mu}, \dots, h_{\ell\mu})$, $(\mu \geq 1)$, such that the sequence

$$n_\mu = \max\{|h_{1\mu}|, \dots, |h_{\ell\mu}|\}, \qquad (\mu \geq 1)$$

is non-decreasing, and define the sequence $\zeta_1, \zeta_2, \dots$ by

$$\zeta_\mu = h_{1\mu}y_1 + \cdots + h_{\ell\mu}\zeta_\ell, \qquad (\mu \geq 1),$$

so that

$$\{\zeta_\mu,\ \mu \geq 1\} = \{h_1y_1 + \cdots + h_\ell\zeta_\ell,\ h = (h_1, \dots, h_\ell) \in \mathbb{Z}^\ell\} = \mathbb{Z}y_1 + \cdots + \mathbb{Z}y_\ell.$$

In fact the poles of any f_i should be removed from this sequence; even more, any point which is too close to a pole should also be omitted, in order to estimate $M_2^{(i)}(Q)$; but, as mentioned above, we give here only a sketch of the proof. Suitable positive constants $c_0, c_1, \dots$ are then selected, which do not depend on the large integer q, so that the Main Theorem can be used with the following inequalities:

$$\begin{gathered}
\partial(q) \leq c_0, \qquad M^{(i)}(R) \leq \exp(c_1 R^{\varrho_i}), \qquad (1 \leq i \leq d),\\
c_2 Q^\ell \leq N(Q) \leq c_3 Q^\ell, \qquad D(Q) \leq c_4 Q,\\
\max\{M_1^{(i)}(Q), M_2^{(i)}(Q)\} \leq \exp(c_5 Q^{\varrho_i}), \qquad (1 \leq i \leq d).
\end{gathered}$$

In the case where $\kappa = 1$, we define $\{\mu_1, \mu_2, \dots\}$ as he sequence of integers μ such that $h_{1\mu} = 0$, and we use the bound

$$N_1(Q) \leq c_6 Q^{\ell-\kappa};$$

in the case $\kappa = 0$, we define $\mu_r = r$, $(r \geq 1)$, and again we have the same upper bound for $N_1(Q)$. Now choose

$$L_i = [c_7 q^{\lambda - \varrho_i}], \quad (1 \leq i \leq d), \quad \text{with} \quad \lambda = (\ell - \kappa + \varrho_1 + \cdots + \varrho_d)/d,$$

and choose also $R = c_{10}Q$, where c_{10} is sufficiently large, so that the desired conclusion follows from the Main Theorem. □

Remark: We need to take for L_i natural integers; we have introduced an integral part, but we need to check that L_i does not vanish; hence the sketch of proof is valid only when

$$\lambda > \varrho^*, \quad \text{where} \quad \varrho^* = \max\{\varrho_1, \dots, \varrho_d\}.$$

This condition can be written

$$\varrho^* \leq \frac{\varrho_1 + \ldots + \varrho_d + \ell - \kappa}{d};$$

since the goal is to prove the upper bound $\ell \leq (\varrho_1 + \cdots + \varrho_d - \kappa)/(d-1)$, we may consider that this condition on ϱ^* is satisfied as soon as

$$\varrho^* \leq \frac{\varrho_1 + \ldots + \varrho_d - \kappa}{d-1};$$

this assumption occurs explicitly in Theorem 1 p. 74 of [36]; however it is possible to remove it, by means of an induction argument on d; see p. 78 of [57] and p. 52 of [58].

The statement of Theorem 1 in [36] also involves a notion of *weighted sequences* which has been used in [57].

c) Corollaries

We first provide a collection of upper bounds for $\delta(f_1, \ldots, f_d)$, where f_i are either linear, or exponential, or elliptic functions; all these estimates follow from Ramachandra's δ–Theorem.

We start with Gel'fond-Schneider's Theorem (already quoted above): *for any non-zero complex number t, $\delta(z, e^{tz}) \leq 1$.*

Another example which do not involve elliptic functions is the six exponentials Theorem (see below): *if $x_1, \ldots, x_d$ are $\mathbb{Q}$-linearly independent complex numbers with $d \geq 2$, then $\delta(e^{x_1 z}, \ldots, e^{x_d z}) \leq d/(d-1)$.* As a matter of fact it suffices to select either $d = 2$ or else $d = 3$ to cover all cases (see below).

The next examples all involve elliptic functions. Notations are as follows: $\wp, \wp^*, \wp_1, \ldots$ are Weierstrass elliptic functions, all of whose invariants $g_2, g_3, g_2^*, g_3^*, \ldots$ are algebraic. The numbers $t, t^*, t_1, \ldots$ are non-zero complex numbers, while $\omega, \omega^*, \omega_1 \ldots$ are respectively non-zero periods of $\wp, \wp^*, \wp_1, \ldots$ Then

$$\begin{array}{rlrl}
\delta(z, \wp(z)) & \leq 2, & & \\
\delta(e^z, \wp(tz)) & \leq 3, & \delta(e^{2\pi i z}, \wp(\omega z)) & \leq 2, \\
\delta(\wp(tz), \wp^*(z)) & \leq 4, & \delta(\wp(\omega z), \wp^*(\omega^* z)) & \leq 3, \\
\delta(e^z, \wp(tz), \wp^*(t^* z)) & \leq 2, & & \\
\delta(\wp_1(t_1 z), \wp_2(t_2 z), \wp_3(z)) & \leq 3, & \delta(\wp_1(\omega_1 z), \wp_2(\omega_2 z), \wp_3(\omega_3 z)) & \leq 2, \\
\delta(\wp_1(t_1 z), \wp_2(t_2 z), \wp_3(t_3 z), \wp_4(z)) & \leq 2. & &
\end{array}$$

We tacitly assumed that the functions we consider are algebraically independent; by the way, it was a non-trivial problem to provide explicit conditions which guarantee the algebraic independence of the functions; an important contribution to this question is Lemma 7 in [36] p. 83; this problem has been solved later in [8].

We now consider more closely a few of these results.

Example 1. $\delta(z, e^{tz})$. – *Hilbert's seventh problem by Schneider's method*

Corollary 1. – *Let α be a non-zero complex algebraic number, and β an irrational algebraic number; choose any determination $\log \alpha$ of the logarithm of α with $\log \alpha \neq 0$ in case $\alpha = 1$. Then $\alpha^\beta = \exp(\beta \log \alpha)$ is a transcendental number.*

Proof: This statement (Theorem of Gel'fond-Schneider) follows from Ramachandra δ–Theorem by taking

$$d = 2, \qquad f_1(z) = z, \quad f_2(z) = \alpha^z = \exp(z \log \alpha), \qquad \varrho_1 = 0, \quad \varrho_2 = 1, \quad \kappa = 0;$$

since $y_1 = 1$ is a common pseudo-algebraic point of f_1 and f_2, we deduce $\delta(f_1, f_2) = 1$, and hence β is not a pseudo-algebraic point of f_2, which means that α^β is a transcendental number. □

Notice that Lang's above mentioned criterion (2) for Schneider's method in [22] does not cover the transcendence of α^β: the point is that the orders of the two functions are 0 and 1 respectively, and it is not sufficient to consider the maximum of both numbers.

Example 2. $\delta(e^z, e^{tz})$. – *The six exponentials Theorem*

The story starts with Ramanujan's study of highly composite numbers; see [1, 20, 21, 22, 36, 37]; see also [2, 58] and [64] for further references.

Corollary 2. – *Let $x_1, \dots, x_d$ be $\mathbb{Q}$-linearly independent complex numbers, and $y_1, \dots, y_\ell$ be also $\mathbb{Q}$-linearly independent complex numbers; assume that the $d\ell$ numbers*

$$e^{x_i y_j}, \qquad (1 \le i \le d, \quad 1 \le j \le \ell)$$

are all algebraic; then $d\ell \le d + \ell$.

Proof: Take

$$f_i(z) = e^{x_i z} \quad \text{and} \quad \varrho_i = 1 \quad \text{for} \quad 1 \le i \le d.$$

□

Example 3. $\delta(e^z, \wp(tz))$

Corollary 3. – *Let $\lambda_1, \dots, \lambda_\ell$ be complex numbers which are linearly independent over $\mathbb{Q}$ such that the ℓ numbers $e^{\lambda_j}, (j = 1, \dots, \ell)$ are algebraic. Let $\wp$ be a Weierstrass elliptic function with algebraic invariants g_2, g_3, and let $v_1, \dots, v_\ell$ be pseudo-algebraic point of $\wp$, not all of which are zero.*
a) If $\ell \ge 4$, then the matrix with 2 rows and ℓ columns

$$\begin{pmatrix} \lambda_1 & \dots & \lambda_\ell \\ u_1 & \dots & u_\ell \end{pmatrix}$$

has rank 2.
b) Assume λ_1 is a rational multiple of $2\pi i$, and u_1 is a period of $\wp$; then the same conclusion holds for $\ell = 3$.
c) If Ramachandra's δ–Conjecture $\delta(e^z, \wp(tz)) \le 1$ holds, the same conclusion is valid already for $\ell \ge 2$.

A nice consequence of part b) of this statement is Corollary p. 87 of [Ramachandra, 1968] which we reproduce here:

If a and b are real positive algebraic numbers different from 1 for which $\log a/\log b$ is irrational and $a < b < a^{-1}$, then one at least of the two numbers

$$x = \left(\frac{1}{240} + \sum_{n=1}^{\infty} \frac{n^3 a^n}{1-a^n}\right) \prod_{n=1}^{\infty} (1-a^n)^{-8},$$

$$y = \left\{\frac{6}{(b^{1/2}-b^{-1/2})^4} - \frac{1}{(b^{1/2}-b^{-1/2})^2} - \sum_{n=1}^{\infty} \frac{n^3 a^n (b^n + b^{-n})}{1-a^n}\right\} \prod_{n=1}^{\infty} (1-a^n)^{-8},$$

is transcendental.

Ramachandra deduces from his results some new transcendental complex numbers, by means of the clean trick (p. 68): *if x and y are real numbers, then the complex number $x+iy$ is transcendental if and only if one at least of the two numbers x, y is transcendental.* Further consequences of this idea have been worked out by G. Diaz in [11] and [12]. See also [68].

Example 4. $\delta(\wp(tz), \wp^*(z))$

We consider now two elliptic functions (compare with [36] p. 68).

Corollary 4. – *Let $\wp$ and $\wp^*$ be two Weierstrass elliptic functions with algebraic invariants g_2, g_3 and g_2^*, g_3^* respectively; let $u_1, \dots, u_\ell$ be $\mathbb{Q}$-linearly independent complex numbers, each of which is a pseudo-algebraic point of $\wp$; similarly, let $u_1^*, \dots, u_\ell^*$ be pseudo-algebraic points of $\wp^*$ which are linearly independent over $\mathbb{Q}$. Assume that the two functions $\wp(u_1 z)$ and $\wp^*(u_1^* z)$ are algebraically independent.*
a) Assume $\ell \geq 5$. Then the matrix

$$\begin{pmatrix} u_1 & \dots & u_\ell \\ u_1^* & \dots & u_\ell^* \end{pmatrix}$$

has rank 2.
b) Assume u_1 is a period of $\wp$ and u_1^* is a period of $\wp^*$; then the rank of the matrix is 2 also when $\ell = 4$.
c) If Ramachandra's δ–Conjecture $\delta(\wp(tz), \wp^*(z)) \leq 1$ holds, the same conclusion is true if only $\ell \geq 2$.

An impressive collection of further consequences to Ramachandra's Main Theorem is displayed in [8] section III C. Several of the previous corollaries deal with elliptic integrals of the first of second kind; further consequences concern elliptic integrals of the third kind as well. More generally, a natural situation where all hypotheses are satisfied is connected with analytic subgroups of commutative algebraic groups (see for instance [57] §3 section 5:"Application du théorème de Ramachandra à l'étude de sous-groupes à un paramètre de certaines variétés de groupes"; see also [7] and [62]); however it seems to the author that this is not exactly the right place to develop this aspect of the theory.

3 Application to density statements

Ramachandra's results on algebraic values of algebraically additive functions can be used to prove some density results. We give here a sample of results dealing with $(\mathbb{R}^\times)^2$,

$\mathbb{C}^\times$, $\mathbb{R}^\times \times E(\mathbb{R})$, $E(\mathbb{R}) \times E^*(\mathbb{R})$ and $E(\mathbb{C})$. Further topological groups are considered in [65].

a1) Consequences of the six exponentials Theorem: real case

Let $\gamma_1, \ldots, \gamma_\ell$ be multiplicatively independent elements in $(\mathbb{R}_+^\times)^2$; write

$$\gamma_j = (\alpha_j, \beta_j), \qquad (j = 1, \ldots, \ell).$$

By means of a well-known result due to Kronecker†, one can show that the subgroup Γ which is generated by $\gamma_1, \ldots, \gamma_\ell$ is dense in $(\mathbf{R}_+^\times)^2$ if and only if for each $s \in \mathbf{Z}^\ell \setminus \{0\}$, the matrix with three rows and ℓ columns

$$\det \begin{pmatrix} \log \alpha_1 & \cdots & \log \alpha_\ell \\ \log \beta_1 & \cdots & \log \beta_\ell \\ s_1 & \cdots & s_\ell \end{pmatrix}$$

has rank 3. An obvious necessary condition is that for all $(a, b) \in \mathbf{Z}^2 \setminus \{0\}$, at least two of the ℓ numbers

$$a \log \alpha_1 + b \log \beta_1, \ldots, a \log \alpha_\ell + b \log \beta_\ell$$

are $\mathbf{Q}$-linearly independent. We assume now that this condition is satisfied, and also that the 2ℓ numbers α_j and β_j are algebraic.

a) According to the six exponentials Theorem, if $\ell \geq 4$, then Γ is dense in $(\mathbf{R}_+^\times)^2$.

For instance if $\ell = 4$ and if the eight numbers α_j, β_j are multiplicatively independent, the group Γ whose rank is 4 is dense in $(\mathbf{R}_+^\times)^2$.

b) If we take for granted the four exponentials Conjecture (see section 2 a), Γ is dense in $(\mathbf{R}_+^\times)^2$ as soon as $\ell \geq 3$.

For instance when $\ell = 3$ and when the six numbers α_j, β_j are multiplicatively independent, then we expect Γ to be dense in $(\mathbf{R}_+^\times)^2$.

Example: *the field* $\mathbf{Q}(\sqrt{2})$.

From the six exponentials Theorem follows that the subgroup of $(\mathbf{R}^\times)^2$, of rank 4, which is generated by the images of

$$2\sqrt{2} - 1, \qquad -3\sqrt{2} - 1, \qquad 4\sqrt{2} - 1, \qquad 6\sqrt{2} - 1,$$

under the canonical embedding of the real quadratic field $\mathbf{Q}(\sqrt{2})$, is dense in $(\mathbf{R}^\times)^2$. If we knew the four exponentials Conjecture we could omit $6\sqrt{2} - 1$ and get a dense subgroup of rank 3.

a2) Consequences of the six exponentials Theorem: complex case

Consider ℓ multiplicatively independent complex numbers $\gamma_1, \ldots, \gamma_\ell$. The subgroup Γ of $\mathbf{C}^\times$ generated by $\gamma_1, \ldots, \gamma_\ell$ is dense if and only if for all $s \in \mathbf{Z}^{\ell+1} \setminus \{0\}$, the matrix with three rows and $\ell + 1$ columns

$$\begin{pmatrix} 0 & \log|\gamma_1| & \cdots & \log|\gamma_\ell| \\ 2\pi i & \log(\gamma_1/\overline{\gamma}_1) & \cdots & \log(\gamma_\ell/\overline{\gamma}_\ell) \\ s_0 & s_1 & \cdots & s_\ell \end{pmatrix}$$

---

†According to his own taste, the reader will find a reference either in
N. Bourbaki, *Eléments de Mathématique*, Topologie Générale, Herman 1974, Chap. VII, § 1, N°1, Prop. 2;
or else in
G.H. Hardy and A.M. Wright, *An Introduction to the Theory of Numbers*, Oxford Sci. Publ., 1938, Chap. XXIII.

has rank 3 (this condition clearly does not depend on the choice of the logarithms $\log(\gamma_j/\overline{\gamma}_j)$). A first necessary condition is that at least two of the numbers $|\gamma_1|, \ldots, |\gamma_\ell|$ are multiplicatively independent; a second necessary condition is that the numbers $\gamma_1/|\gamma_1|, \ldots, \gamma_\ell/|\gamma_\ell|$ are not all roots of unity. These two conditions mean that the projection of Γ on each of the two factors $\mathbf{R}_+^\times$ and $\mathbf{R}/\mathbf{Z}$ which arises from $z \mapsto (|z|, z/|z|)$, has a dense image. We assume that these conditions are satisfied, and furthermore that the ℓ complex numbers γ_j are algebraic. According to the four exponentials Conjecture, Γ should be dense in $\mathbf{C}^\times$ without any further assumption. On the other hand, if we use the six exponentials Theorem, assuming that three at least of the numbers $|\gamma_1|, \ldots, |\gamma_\ell|$ are multiplicatively independent, we deduce that Γ is dense in $\mathbf{C}^\times$.

Example: the field $\mathbf{Q}(i)$.

The subgroup of rank 3 in $\mathbf{C}^\times$ which is generated by

$$2+i, \qquad 2+3i, \qquad 4+i,$$

is dense $\mathbf{C}^\times$. If the four exponential Conjecture holds, then for instance the subgroup of rank 2

$$\{(2+i)^s 2^t;\ (s,t) \in \mathbf{Z}^2\},$$

generated by 2 and $2+i$ is dense in $\mathbf{C}^\times$; a proof of this result would follow from a special case of the four exponentials Conjecture: it would suffice to show that the three numbers

$$\log 2, \quad \log 5, \quad \frac{\log 2}{2\pi i} \cdot \log\left(\frac{3+4i}{5}\right)$$

are $\mathbf{Q}$-linearly independent, which means that for each $(\lambda, \mu) \in \mathbf{Q}^2$, the determinant

$$\det\begin{pmatrix} \log 5 & \log 2 \\ \log\left(\dfrac{3+4i}{5}\right) + 2\lambda\pi i & 2\mu\pi i \end{pmatrix}$$

does not vanish. This is not yet known.

b) Product of the multiplicative group with an elliptic curve

Let $\wp$ be a Weierstrass elliptic function with real algebraic invariants g_2, g_3:

$$\wp'^2 = 4\wp^3 - g_2\wp - g_3;$$

the set of real points on the corresponding elliptic curve E, namely

$$E(\mathbf{R}) = \{(x:y:t) \in \mathbf{P}_2(\mathbf{R});\ y^2t = 4x^3 - g_2xt^2 - g_3t^3\},$$

has one or two connected components, according as the polynomial $4X^3 - g_2X - g_3$ has one or three real roots; we denote by $E(\mathbf{R})^0$ the connected component of $E(\mathbf{R})$ which contains the origin $(0:1:0)$; hence $E(\mathbf{R})^0$ is a subgroup of $E(\mathbf{R})$ of index 1 or 2. For simplicity of notation, when ω is a pole of $\wp$, then $(\wp(\omega):\wp'(\omega):1)$ means $(0:1:0)$. With this convention the map $\exp_E: u \mapsto (\wp(u):\wp'(u):1)$ is a surjective homomorphism from $\mathbf{R}$ onto $E(\mathbf{R})^0$ whose kernel is of the form $\mathbf{Z}\omega$, where ω is a fundamental real period of $\wp$. A point $\gamma = \exp_E(u) = (\wp(u):\wp'(u):1) \in E(\mathbf{R})^0$ is a torsion point if and only if the two number u and ω are linearly dependent over $\mathbf{Q}$. More generally, when $u_1, \ldots, u_\ell$ are real numbers, the rank over $\mathbf{Z}$ of the subgroup Γ generated by the ℓ points $\gamma_j = (\wp(u_j) : \wp'(u_j) : 1)$, $(1 \le j \le \ell)$ in $E(\mathbf{R})^0$ is related to the rank of the subgroup Y of $\mathbf{R}$ generated by the $\ell + 1$ real numbers $u_1, \ldots, u_\ell, \omega$ by $\mathrm{rank}_{\mathbf{Z}}\,\Gamma = \mathrm{rank}_{\mathbf{Z}}\,Y - 1$.

Recall (Kronecker's Theorem again) that a subgroup of rank ≥ 1 in $\mathbf{R}/\mathbf{Z}$ is dense. We deduce:

> *Let E be an elliptic curve which is defined over $\mathbf{R}$ and let γ be a point of infinite order on $E(\mathbf{R})^0$; then the subgroup $\mathbf{Z}\gamma$ is dense for the real topology in $E(\mathbf{R})^0$.*

The next density result deals with the product of the multiplicative group of non-zero real numbers with $E(\mathbf{R})$; it will be proved as a consequence of Corollary 3.

Corollary 5. – *Let $\alpha_1, \ldots, \alpha_\ell$ be multiplicatively independent positive real algebraic numbers; let $\gamma_1, \ldots, \gamma_\ell$ be points on $E(\overline{\mathbf{Q}}) \cap E(\mathbf{R})^0$, which are not all torsion points. Denote by Γ the subgroup of $\mathbf{R}_+^\times \times E(\mathbf{R})^0$ which is spanned by the ℓ points (α_j, γ_j), $(1 \le j \le \ell)$.*

a) Assume that $\ell \ge 4$. Then Γ is dense in $\mathbf{R}_+^\times \times E(\mathbf{R})^0$.
b) According to Ramachandra's δ–Conjecture $\delta(e^z, \wp(tz)) \le 1$, the same conclusion should hold as soon as $\ell \ge 2$.

Proof: The exponential map

$$\begin{array}{ccc} \mathbf{R}^2 & \longrightarrow & \mathbf{R}_+^\times \times E(\mathbf{R})^0 \\ (x_1, x_2) & \longmapsto & \big(e^{x_1}; (\wp(x_2) : \wp'(x_2) : 1)\big) \end{array}$$

is a topological surjective homomorphism with kernel $\mathbf{Z}(0, \omega)$ for some $\omega \in \mathbf{R}^\times$; define $y_0 = (0, \omega)$ and $y_j = (\log \alpha_j, u_j)$, $(1 \le j \le \ell)$, where $u_j \in \mathbf{R}$ is such that $\gamma_j = (\wp(u_j) : \wp'(u_j) : 1)$. Now the goal is to prove that the subgroup $Y = \mathbf{Z}y_0 + \mathbf{Z}y_1 + \cdots + \mathbf{Z}y_\ell$ is dense in $\mathbf{R}^2$. Let $\varphi : \mathbf{R}^2 \to \mathbf{R}$ be a linear form which satisfies $\varphi(Y) \subset \mathbf{Z}$. According to Kronecker's above mentioned Theorem (see footnote ([1])), we only need to prove $\varphi = 0$. Write $\varphi(y_j) = s_j$ with $s_j \in \mathbf{Z}$, $(0 \le j \le \ell)$; then the matrix

$$\begin{pmatrix} 0 & \log \alpha_1 & \cdots & \log \alpha_\ell \\ \omega & u_1 & \cdots & u_\ell \\ s_0 & s_1 & \cdots & s_\ell \end{pmatrix}$$

has rank < 3; it follows that the rank of the matrix

$$\begin{pmatrix} s_0 \log \alpha_1 & \cdots & s_0 \log \alpha_\ell \\ s_0 u_1 - s_1 \omega & \cdots & s_0 u_\ell - s_\ell \omega \end{pmatrix}$$

is < 2. Since $u_1, \ldots, u_\ell$ are not all torsion points and $\log \alpha_1, \ldots, \log \alpha_\ell$ are linearly independent over $\mathbf{Q}$, it follows from Corollary 3 that $s_0 = 0$; using once more the linear independence of the $\log \alpha$'s, we deduce $s_1 = \cdots = s_\ell = 0$, and $\varphi = 0$. □

Remark: Ramachandra's δ–Conjecture $\delta(e^z, \wp(tz)) \le 1$ implies the following:

> *Let $\log \alpha_1$ and $\log \alpha_2$ be two $\mathbf{Q}$-linearly independent logarithms of algebraic numbers. Let E be an elliptic curve which is defined over the field $\overline{\mathbf{Q}}$ of algebraic numbers, ω*

be a non-zero period of $\exp_E$ *and* $u \in \mathbf{C}$ *be an elliptic logarithm of a point of infinite order in* $E(\overline{\mathbf{Q}})$. *Then the three numbers*

$$\frac{\log \alpha_1}{\log \alpha_2}, \quad \frac{u}{\omega}, \quad 1$$

are linearly independent over $\mathbf{Q}$.

Indeed, this means, for each $(\lambda, \mu) \in \mathbf{Q}^2$,

$$\det \begin{pmatrix} \log \alpha_1 & \log \alpha_2 \\ u + \lambda\omega & \mu\omega \end{pmatrix} \neq 0.$$

c) Product of two elliptic curves

As a consequence of Corollary 4 we have a density result for the product of two elliptic curves over the real number field.

Corollary 6. – *Let* E *and* E^* *be two Weierstrass elliptic curves with real algebraic invariants* g_2, g_3 *and* g_2^*, g_3^* *respectively; assume for simplicity that there is no isogeny between them, which means that for* $t \in \mathbf{C}^\times$, *the two functions* $\wp(tz)$ *and* $\wp^*(z)$ *are algebraically independent. Denote by* ω *(resp.* ω^**) a non-zero real period of* $\wp$ *(resp. of* $\wp^*$*). Let* ℓ *be a positive integer and let* $\gamma_1, \dots, \gamma_\ell$ *(resp.* $\gamma_1^*, \dots, \gamma_\ell^*$*) be elements in* $E(\mathbf{R})^0 \cap E(\overline{\mathbf{Q}})$ *(resp. in* $E^*(\mathbf{R})^0 \cap E^*(\overline{\mathbf{Q}})$*) such that*

a) $\gamma_1, \dots, \gamma_\ell$ *are not all torsion points on* $E(\overline{\mathbf{Q}})$;
b) $\gamma_1^*, \dots, \gamma_\ell^*$ *are not all torsion points on* $E^*(\overline{\mathbf{Q}})$;
c) the subgroup Γ *of* $E(\overline{\mathbf{Q}}) \times E^*(\overline{\mathbf{Q}})$ *generated by the* ℓ *points* (γ_j, γ_j^*), $(1 \leq j \leq \ell)$ *has rank* ℓ.
Then
1) if $\ell \geq 3$, Γ *is dense in* $E(\mathbf{R})^0 \times E^*(\mathbf{R})^0$.
2) if Ramachandra's δ*–Conjecture* $\delta(\wp(\omega z), \wp^*(\omega^* z)) \leq 1$ *is true, the same conclusion holds already for* $\ell \geq 1$.

Proof: We first translate the hypotheses concerning the points on the elliptic curves in terms of elliptic logarithms. Let ω (resp. ω^*) be a real fundamental period of $\wp$ (resp. of $\wp^*$); for $1 \leq j \leq \ell$, let $u_j \in \mathbf{R}$ satisfy $(\wp(u_j) : \wp'(u_j) : 1) = \gamma_j$, and let $u_j^* \in \mathbf{R}$ satisfy $(\wp^*(u_j^*) : \wp^{*\prime}(u_j^*) : 1) = \gamma_j^*$. Now $u_1, \dots, u_\ell$ (resp. $u_1^*, \dots, u_\ell^*$) are pseudo-algebraic points of $\wp$ (resp. of $\wp^*$), such that
a) two at least of the $\ell + 1$ numbers $u_1, \dots, u_\ell, \omega$ are linearly independent over $\mathbf{Q}$,
b) two at least of the $\ell + 1$ numbers $u_1^*, \dots, u_\ell^*, \omega^*$ are linearly independent over $\mathbf{Q}$,
c) the $\ell + 2$ points

$$y_0 = \begin{pmatrix} \omega \\ 0 \end{pmatrix}, \qquad y_0^* = \begin{pmatrix} 0 \\ \omega^* \end{pmatrix}, \qquad y_j = \begin{pmatrix} u_j \\ u_j^* \end{pmatrix}, \quad (1 \leq j \leq \ell)$$

are linearly independent over $\mathbf{Q}$. We want to prove that the subgroup of $\mathbf{R}^2$ of rank $\ell + 2$ which is generated by $y_0, y_0^*, y_1, \dots, y_\ell$ is dense in $\mathbf{R}^2$. Indeed, let $s_0, s_0^*, s_1, \dots, s_\ell$ be rational integers, not all of which are zero; we shall deduce from Corollary 4 (with a shift of notations $\ell \mapsto \ell + 1$) that the matrix

$$\begin{pmatrix} \omega & 0 & u_1 & \cdots & u_\ell \\ 0 & \omega^* & u_1^* & \cdots & u_\ell^* \\ s_0 & s_0^* & s_1 & \cdots & s_\ell \end{pmatrix}$$

has rank 3. If $s_0 = 0$ (resp if $s_0^* = 0$), this follows from the hypothesis b) (resp. a)) above. Assume now $s_0 \neq 0$ and $s_0^* \neq 0$; we want to prove that the matrix

$$\begin{pmatrix} -s_0^*\omega & s_0u_1 - s_1\omega & \cdots & s_0u_\ell - s_\ell\omega \\ \omega^* & u_1^* & \cdots & u_\ell^* \end{pmatrix}$$

has rank 2; in order to use part b) of Corollary 4, we need to check that the $\ell + 1$ elements on the first row are linearly independent over $\mathbf{Q}$, and the same for the $\ell + 1$ elements on the second row; if this were not the case, and if the matrix had rank < 2, we would get a non trivial linear dependence relation between the $\ell + 2$ elements

$$\begin{pmatrix} \omega \\ 0 \end{pmatrix}, \quad \begin{pmatrix} 0 \\ \omega^* \end{pmatrix}, \quad \begin{pmatrix} u_j \\ u_j^* \end{pmatrix}, \quad (1 \leq j \leq \ell),$$

which would contradict the assumption $\mathrm{rank}_{\mathbf{Z}}\,\Gamma = \ell$. □

Remark: Ramachandra's δ–Conjecture $\delta(\wp(\omega z), \wp^*(\omega^* z)) \leq 1$ implies the following:

> *Let $\wp$ (resp. $\wp^*$) be a Weierstrass elliptic function with algebraic g_2, g_3 (resp. g_2^*, g_3^*); let ω (resp. ω^*) be a non-zero period of $\wp$ (resp. of $\wp^*$), and $u \in \mathbf{C}$ (resp. $u^* \in \mathbf{C}$) be a pseudo-algebraic point of $\wp$ (resp. of $\wp^*$), with u/ω and u^*/ω^* both irrational numbers; assume that the two complex functions $\wp(\omega z)$ and $\wp^*(\omega^* z)$ are algebraically independent; then the three numbers*
>
> $$1, \ \frac{u}{\omega}, \ \frac{u^*}{\omega^*}$$
>
> *are linearly independent over* $\mathbf{Q}$.

In other words, according to this conjecture, for rational integers s_0, s_0^*, s, the determinant

$$\det\begin{pmatrix} \omega & 0 & u \\ 0 & \omega^* & u^* \\ s_0 & s_0^* & s \end{pmatrix}$$

can vanish only for $s_0 = s_0^* = s = 0$.

d) Complex points on an elliptic curve

Let E be an elliptic curve over $\mathbf{C}$

$$E(\mathbf{C}) = \{(x : y : t) \in \mathbf{P}_2(\mathbf{C});\ y^2t = 4x^3 - g_2xt^2 - g_3t^3\},$$

and let $\gamma \in E(\mathbf{C})$; we ask whether the subgroup $\mathbf{Z}\gamma$ spanned by γ is dense in the topological group $E(\mathbf{C})$.

Denote as before by $\wp$ the Weierstrass elliptic function with invariants g_2 and g_3:

$$\wp'^2 = 4\wp^3 - g_2\wp - g_3,$$

by $\Omega = \mathbf{Z}\omega_1 + \mathbf{Z}\omega_2$ the lattice of periods of $\wp$, by $\overline{\Omega} = \mathbf{Z}\overline{\omega}_1 + \mathbf{Z}\overline{\omega}_2$ the complex conjugate lattice and by $\overline{E} = \mathbf{C}/\overline{\Omega}$ the Weierstrass elliptic curve with invariants $\overline{g}_2$ and $\overline{g}_3$; select $u = x_1\omega_1 + x_2\omega_2 \in \mathbf{C}$ (with real x_1 and x_2) such that $\gamma = (\wp(u) : \wp'(u) : 1)$. The three conditions

(i) γ is not a torsion point;
(ii) the three numbers u, ω_1, ω_2 are $\mathbf{Q}$-linearly independent;
(iii) $(x_1, x_2) \notin \mathbf{Q}^2$
are equivalent.

On the other hand the three following conditions are also equivalent:
(j) $\mathbf{Z}\gamma$ is dense in the topological group $E(\mathbf{C})$;
(jj) the three numbers $1, x_1, x_2$ are $\mathbf{Q}$-linearly independent over $\mathbf{Q}$;
(jjj) if $\omega \in \Omega$ is any non-zero period of $\wp$ and $n \geq 1$ any positive integer, then $n\gamma$ does not belong to the 1-parameter subgroup $\{(\wp(t\omega) : \wp'(t\omega) : 1);\ t \in \mathbf{R}\}$ of $E(\mathbf{C})$.

Of course conditions (j), (jj) et (jjj) imply conditions (i), (ii) et (iii); clearly the converse does not hold without any further assumption. Let us assume that g_2 and g_3 are algebraic, as well as $\wp(u)$ and $\wp'(u)$.

The curve E is defined over $\mathbf{R}$ if and only if there exists $\theta \in \mathbf{C}^\times$ such that $\theta\Omega = \overline{\theta\Omega}$; the set

$$\mathcal{E}(E) = \{\theta \in \mathbf{C}^\times;\ \mathrm{rank}_{\mathbf{Z}}(\theta\Omega \cap \overline{\theta\Omega}) = 2\};$$

is empty if and only if the two curves E and $\overline{E}$ are not isogeneous. We start with the easiest case:

Corollary 7. – *Let E be an elliptic curve which is defined over the field $\overline{\mathbf{Q}}$ of algebraic numbers and is not isogeneous to its complex conjugate.*
1) Any subgroup of $E(\overline{\mathbf{Q}})$ of rank ≥ 3 is dense in $E(\mathbf{C})$ for the complex topology.
2) If Ramachandra's δ–Conjecture $\delta(\wp(\omega z), \wp^*(\omega^* z)) \leq 1$ is true, any element in $E(\overline{\mathbf{Q}})$ which is not a torsion point spans a dense subgroup of $E(\mathbf{C})$.

Proof: The proof is the same as for Corollary 6; also, Corollary 7 will follow from Corollary 8 below. □

Before we study the general case, we prove the following auxiliary lemma.

Lemma 1. – *Let $\Omega = \mathbf{Z}\omega_1 + \mathbf{Z}\omega_2$ be a lattice in $\mathbf{C}$; let $\theta \in \mathbf{C}^\times$ be such that $\theta\Omega \cap \overline{\theta\Omega}$ is a subgroup of finite index in $\theta\Omega$. Let Y be a finitely generated subgroup of $\mathbf{C}$. Define two subgroups of $\mathbf{C}$ by*

$$Y_\theta = \{\theta y - \overline{\theta y};\ y \in Y\}$$

and

$$\widetilde{\Omega}_\theta = \{\theta\omega - \overline{\theta\omega'};\ (\omega, \omega') \in \Omega \times \Omega\} \subset \mathbf{C}.$$

If $Y_\theta \cap \widetilde{\Omega}_\theta$ is a subgroup of finite index in Y_θ, then Y is not dense in $\mathbf{C}$.

Proof: a) From the hypotheses we deduce that there exists a positive integer m such that $m\theta\omega_1 \in \overline{\theta\Omega}$ and $m\theta\omega_2 \in \overline{\theta\Omega}$. Define a, b, c, d is $\mathbf{Z}$ by

$$m\theta\omega_1 = a\overline{\theta\omega_1} + b\overline{\theta\omega_2} \qquad \text{and} \qquad m\theta\omega_2 = c\overline{\theta\omega_1} + d\overline{\theta\omega_2}.$$

We show the relations

$$a + d = 0 \qquad \textit{and} \qquad m^2 + ad - bc = 0.$$

Using complex conjugation, we get

$$m^2\overline{\theta\omega_1} = am\theta\omega_1 + bm\theta\omega_2 = a(a\overline{\theta\omega_1} + b\overline{\theta\omega_2}) + b(c\overline{\theta\omega_1} + d\overline{\theta\omega_2})$$

and

$$m^2\overline{\theta\omega_2} = cm\theta\omega_1 + dm\theta\omega_2 = c(a\overline{\theta\omega_1} + b\overline{\theta\omega_2}) + d(c\overline{\theta\omega_1} + d\overline{\theta\omega_2}),$$

hence

$$m^2 = a^2 + bc, \quad (a+d)b = 0, \qquad m^2 = d^2 + bc, \quad (a+d)c = 0.$$

The solution $a = d$ and $b = c = 0$ is not possible because $\omega_2/\omega_1 \notin \mathbf{R}$.

b) *We show that there exist ω_0 and ω_0' which generate a subgroup of finite index in Ω such that $\theta\omega_0 \in \mathbf{R}$ and $\theta\omega_0' \in i\mathbf{R}$.*

We want to find λ, μ in $\mathbf{Z}$ such that $\omega_0 = \lambda\omega_1 + \mu\omega_2$ satisfies $\theta\omega_0 = \overline{\theta\omega_0}$: we need to solve the system

$$(a-m)\lambda + c\mu = 0$$
$$b\lambda + (d-m)\mu = 0$$

whose determinant $ad - bc - m(a+d) + m^2$ vanishes; hence there is a non trivial solution, which means that $\theta\Omega\cap\mathbf{R}$ is a $\mathbf{Z}$-module of rank 1. Similarly, since $ad-bc+m(a+d)+m^2 = 0$, the system

$$(a+m)\lambda' + c\mu' = 0$$
$$b\lambda' + (d+m)\mu' = 0$$

has a non trivial solution $(\lambda', \mu') \in \mathbf{Z}^2$, and $\theta\Omega \cap i\mathbf{R}$ is generated by a non-zero element $\omega_0' = \lambda'\omega_1 + \mu'\omega_2$.

c) *For each $v \in \widetilde{\Omega}_\theta \cap i\mathbf{R}$ we show that there exist a positive integer k and an element $\omega \in \Omega$ such that*

$$kv = \theta\omega - \overline{\theta\omega}.$$

Since $v \in \widetilde{\Omega}_\theta$ there exist $\omega \in \Omega$ and $\omega' \in \Omega$ such that $v = \theta\omega - \overline{\theta\omega'}$. It follows from the previous result in b) that there exist integers h, a, b, c, d with $h \geq 1$ such that

$$h\omega = a\omega_0 + b\omega_0' \qquad \text{and} \qquad h\omega' = c\omega_0 + d\omega_0'.$$

We deduce

$$hv = (a-c)\theta\omega_0 + (b+d)\theta\omega_0'.$$

From the hypothesis $\overline{v} = -v$ we conclude $a = c$, and the result follows with $k = 2h$ and $\omega = (b+d)\omega_0'$.

d) *Define a linear form $\varphi : \mathbf{C} \to \mathbf{R}$ of $\mathbf{R}$-vector spaces by*

$$\varphi(x_1\omega_1 + x_2\omega_2) = \mu x_1 - \lambda x_2,$$

where (λ, μ) satisfies (as before) $\omega_0 = \lambda\omega_1 + \mu\omega_2 \in \Omega \cap \mathbf{R}$. If $Y_\theta \cap \widetilde{\Omega}_\theta$ is a subgroup of finite index in Y_θ, then we shall deduce $\varphi(Y) \subset \mathbf{Q}$ (from which it follows that Y is not dense in $\mathbf{C}$).

For the proof, we take $y = x_1\omega_1 + x_2\omega_2 \in Y$ with $(x_1, x_2) \in \mathbf{R}^2$. Let m be a positive integer such that $m(\theta y - \overline{\theta y}) \in \widetilde{\Omega}_\theta$; it follows from c) that there exist $k \geq 1$ and $\omega \in \Omega$ with

$$mk(\theta y - \overline{\theta y}) = \theta\omega - \overline{\theta\omega},$$

and therefore $mk\theta(y-\omega) \in \mathbf{R}$. Put $n = mk$ and write $\omega = a\omega_1 + b\omega_2$:

$$(nx_1 - a)\theta\omega_1 + (nx_2 - b)\theta\omega_2 \in \mathbf{R};$$

however $\theta\omega_1$ and $\theta\omega_2$ are linearly independent over $\mathbf{R}$ and satisfy $\lambda\theta\omega_1 + \mu\theta\omega_2 \in \mathbf{R}$. We deduce $\lambda(nx_2 - b) = \mu(nx_1 - a)$, which completes the proof. □

We can now state and prove the following result:

Corollary 8. – *Let $E = \mathbf{C}/\Omega$ be a Weierstrass elliptic curve with algebraic g_2, g_3. Define*

$$\mathcal{E}(E) = \{\theta \in \mathbf{C}^\times;\ \mathrm{rank}_{\mathbf{Z}}(\theta\Omega \cap \overline{\theta\Omega}) = 2\};$$

for each $\theta \in \mathcal{E}(E)$, define

$$\widetilde{\Omega}_\theta = \{\theta\omega - \overline{\theta\omega'};\ (\omega, \omega') \in \Omega \times \Omega\} \subset \mathbf{C}.$$

Let $\Gamma = \mathbf{Z}\gamma_1 + \cdots + \mathbf{Z}\Gamma_\ell$ be a finitely generated subgroup of rank ℓ in $E(\overline{\mathbf{Q}})$. Define $Y \subset \mathbf{C}$ by $Y = \exp_E^{-1}(\Gamma)$. For each $\theta \in \mathcal{E}(E)$, put

$$Y_\theta = \{\theta y - \overline{\theta y};\ y \in Y\}.$$

Assume that for each $\theta \in \mathcal{E}(E)$, the subgroup $Y_\theta \cap \widetilde{\Omega}_\theta$ is not of finite index in Y_θ.
a) If $\ell \geq 3$, then Γ is dense in $E(\mathbf{C})$.
b) If Ramachandra's δ–Conjecture $\delta(\wp(\omega z), \wp^*(\omega^* z)) \leq 1$ is true, then Γ is a dense subgroup of $E(\mathbf{C})$.

It follows that for any elliptic curve E which is defined over $\overline{\mathbf{Q}}$, there exists an algebraic number field K such that $E(K)$ is dense in the topological group $E(\mathbf{C})$.

Proof: A necessary and sufficient condition for Γ to be dense in $E(\mathbf{C})$ is that Y is dense in $\mathbf{C}$. For $1 \leq j \leq \ell$, let $u_j \in \mathbf{C}$ be an elliptic logarithm of γ_j; then Y is dense in $\mathbf{C}$ if and only if $\mathbf{Z}\omega_1 + \mathbf{Z}\omega_2 + \mathbf{Z}u_1 + \cdots + \mathbf{Z}u_\ell$ is dense in $\mathbf{C}$; according to Kronecker's Theorem, this is equivalent to the following assertion: for each $(s_0', s_0'', s_1, \ldots, s_\ell) \in \mathbf{Z}^{\ell+2} \setminus \{0\}$, the matrix

$$\begin{pmatrix} \omega_1 & \omega_2 & u_1 & \cdots & u_\ell \\ \overline{\omega_1} & \overline{\omega_2} & \overline{u_1} & \cdots & \overline{u_\ell} \\ s_0' & s_0'' & s_1 & \cdots & s_\ell \end{pmatrix}$$

has rank 3. This condition is clearly satisfied if either $s_0' = 0$ or $s_0'' = 0$, because we assume Γ has rank ℓ. If $s_0' \neq 0$ and $s_0' \neq 0$, we define

$$\omega = s_0'\omega_2 - s_0''\omega_1, \qquad v_j = s_0'u_j - s_j\omega_1, \quad (1 \leq j \leq \ell),$$

and we want to prove that the matrix

$$\begin{pmatrix} \omega & v_1 & \cdots & v_\ell \\ \overline{\omega} & \overline{v_1} & \cdots & \overline{v_\ell} \end{pmatrix}$$

has rank 2. If the two functions $\wp(\omega z)$ and $\overline{\wp}(\overline{\omega}z)$ are algebraically independent, we can apply parts b) and c) of Corollary 4 (with ℓ replaced by $\ell + 1$ again). Otherwise, since the period lattices of $\wp(\omega z)$ and $\overline{\wp}(\overline{\omega}z)$ are respectively $(1/\omega)\Omega$ and $(1/\overline{\omega})\overline{\Omega}$, we can use our assumption on Y_θ with $\theta = 1/\omega$: firstly $\theta\Omega \cap \overline{\theta\Omega}$ is of finite index in $\theta\Omega$, secondly the numbers

$$s_0'(\theta u_1 - \overline{\theta u_1}) - s_1(\theta\omega_1 - \overline{\theta\omega_1})$$

do not all vanish; hence the above matrix has rank 2. □

4 Further contributions of Ramachandra to transcendental number theory

a) On the numbers 2^{π^k}, $(k = 1, 2, 3, \ldots)$

From the six exponentials Theorem follows that one at least of the three numbers 2^{π}, 2^{π^2} and 2^{π^3} is transcendental. The result can be made effective, and a transcendence measure for at least one of these three numbers can be derived. Using a Theorem of Szemeredi, Srinivasan obtained a result which he himself states as follows:

for almost all k, the number 2^{π^k} has a transcendence measure of the type

$$\left|2^{\pi^k} - \alpha\right| \geq \exp\{-(\log H_\nu)^{1+\epsilon}\}$$

(for any $\epsilon > 0$ with respect to a sequence of heights $H_\nu \to \infty$, with height of α bounded by H_ν).

Starting from such a statement, he investigated the number of algebraic numbers among the numbers 2^{π^k}, $(1 \leq k \leq N)$; in [53] and [54] he got the upper bound $O(\sqrt{N})$ (conjecturally, none of them is algebraic). The O constant was bounded by 2 in [42] and by $\sqrt{2}$ in [4]. In fact, $O(\sqrt{N})$ can be replaced by a more explicit expression of the type $c\sqrt{N}$ + smaller order terms.

b) A note on Baker's method

Ramachandra has several contributions to Baker's theory on linear forms in logarithms and its applications; see in particular [38, 39], and [41]. A survey of this subject is given in [3]. It is fair to quote here also the important work of Shorey: [50, 51] and [52] (*one of the main contributions of Ramachandra's to transcendental number theory is Shorey*).

For his investigations concerning lower bounds for linear forms in logarithms of algebraic numbers, Ramachandra used Baker's method (which is a generalization of Gel'fond's solution to Hilbert's seventh problem). It turns out that a method closely related to [36] yields similar results; see [31, 64] and [27] for "usual" logarithms, and [69] for elliptic logarithms.

The paper [38] provides the first lower bound for simultaneous linear forms in logarithms. This was done at a very early stage of Baker's method, and the estimate has been superseded now, but the interest lies in the idea of improving the bound by considering several simultaneous linear forms; also the suggestion that the simultaneous result should allow one to conjecture a stronger result for a single linear form turned out to be correct.

The subject has been developed more recently in [28, 35] and [18]. Using the same argument as in [38], one might consider that these work give partial evidence towards the Lang-Waldschmidt Conjectures (see the Introduction to Chapters X and XI of [23]).

c) An easy transcendence measure for e

When θ is a complex transcendental number, a *transcendence measure* for θ is a lower bound for $|P(\theta)|$ when P is a non-zero polynomial with rational integer coefficients. Such a lower bound should depend on the degree of the polynomial P as well as on the height $H(P)$ of the same (we consider the so-called "usual height", namely the maximum absolute value of the coefficients of P). The very first transcendence measure goes back to the nineteenth century: Borel gave a transcendence measure for e in 1899. Later Popken (1929), Mahler

(1932), Fel'dman (1963), Galochkin (1972), Cijsouw (1974), Durand (1980), Ramachandra (1987), Khassa and Srinivasan (1991) as well as other authors gave further transcendence measure for the same number e. We quote here the result of [19] which rests on Ramachandra's method in [40].

For every positive integer n there exists a constant H_0 which depends only on n such that for each positive integer m, each positive real number $H \geq H_0$ and each non-zero polynomial $P \in \mathbf{Z}[X]$ of degree $\leq n$, with at most m non-zero coefficients, and with height $H(P) \leq H$,

$$|P(e)| \geq H^{-m-\frac{cmn\log(m+1)}{\log\log H}}.$$

An interesting feature of this statement is that it takes into account the number of non-zero coefficients of the polynomial in place of the degree; such an idea appears for instance in some works dealing with complexity theory; in Diophantine approximation it occurs in papers connected with Lehmer's Conjecture; it seems it never occurred before in connection with transcendence measures.

5 Open problems

We have already seen a few unsolved questions, notably the four exponentials Conjecture and Ramachandra's δ–Conjecture. Here are further unsolved questions.

a) Algebraic independence

Assume $x_1, \ldots, x_d$ are $\mathbf{Q}$-linearly independent complex numbers, and y_1, y_2, y_3 are $\mathbf{Q}$-linearly independent complex numbers; the six exponentials Theorem shows that at least $d-1$ of the numbers

$$e^{x_i y_j}, \qquad (1 \leq i \leq d,\ j = 1, 2, 3)$$

are transcendental. A natural question is to ask whether *at least $d-1$ of these numbers are algebraically independent.* This amounts to ask if *the transcendence degree of the field generated by these $3d$ numbers is at least $d-1$.* This problem was raised in [36]. The same argument was reproduced in [56] as follows: according to the four exponentials Conjecture, at least $d-1$ of the numbers

$$e^{x_i y_j}, \qquad (1 \leq i \leq d,\ j = 1, 2)$$

are expected to be transcendental; *is-it true that at least $d-1$ of these numbers are algebraically independent?*

Surprisingly enough, the answer to these questions is no! For instance take $y_1 = 1$, $y_2 = \beta$, $y_3 = \beta^2$, where β is cubic (resp. $y_1 = 1$, $y_2 = \beta$, where β is quadratic) if one wishes to use the six exponentials Theorem (resp. the four exponentials Conjecture), and choose

$$d = 3k, \quad \{x_1, \ldots, x_d\} = \{\log\alpha_j, \beta\log\alpha_j, \beta^2\log\alpha_j;\ (1 \leq j \leq k)\}$$

(resp.

$$d = 2k, \quad \{x_1, \ldots, x_d\} = \{\log\alpha_j, \beta\log\alpha_j;\ (1 \leq j \leq k)\}),$$

where $\alpha_1, \ldots, \alpha_k$ are multiplicatively independent algebraic numbers; the transcendence degree is at most $2k = 2d/3$ (resp. $k = d/2$).

One way of repairing the conjecture (see [58] conjecture 7.5.5 and exercice 7.5.b) is to assume that y_1, y_2 are linearly independent over the field of algebraic numbers. A better view of looking at this kind of problem from a conjectural point of view is to consider Schanuel's Conjecture [22] Chapter 3 p. 30.

The first results of algebraic independence in this direction are due to Gel'fond [15]: if $\ell d \geq 2(\ell + d)$, then the transcendence degree t of the field generated by the $d\ell$ numbers

$$e^{x_i y_j}, \qquad (1 \leq i \leq d,\ 1 \leq j \leq \ell)$$

is at least 2. Gel'fond's statement involved a so-called "technical hypothesis" (measure of linear independence for the x_i's, and also for the y_j's), which was removed later by Tijdeman (see [58] Chapitre 7).

In the early 70's, W.D.Brownawell and A.A.Smelev succeeded to prove $t \geq 3$ under suitable assumptions; in 1974, Chudnovsky obtained $2^t \geq \ell d/(\ell + d)$; references are given in [9]. P.Philippon reached the estimate $t + 1 \geq \ell d/(\ell + d)$ in [33]; this was improved by G. Diaz in [10] as $t \geq [\ell d/(\ell + d)]$ provided that $\ell d > \ell + d$ (without the proviso, the four exponentials Conjecture would follow!). An interesting fact is that all these results on "large transcendence degree" always involve a "technical hypothesis"; it is an open problem to remove it.

The above mentioned theorems of algebraic independence deal with the usual exponential function; here again, extensions can be given to commutative algebraic groups; we only quote [8] and [61] which include extensions to results of algebraic independence of Ramachandra's transcendence results concerning the exponential and elliptic functions.

b) Schneider's second problem in [49]

Ramachandra's method might be the right way towards a solution of the second of Schneider's eight problems in [49]: to prove Schneider's Theorem on the transcendence of $j(\tau)$ for τ an algebraic number in the upper half plane by means of the modular function (and not by mean of elliptic functions). The best known results in this direction are in [55].

c) Linear independence of elliptic logarithms in the non CM case by Schneider's method

We already quoted the paper [69] where lower bounds for linear forms in elliptic logarithms are provided, by means of a method which is closely related to [36]; as a matter of fact, an assumption is needed: namely one assumes that the elliptic curves has non trivial endomorphisms (CM case). It is not known how to extend the method to the non-CM case.

d) Effective results

Ramachandra was concerned (see top of p. 67 in [36]) by the fact that his simplification of Schneider's method might be at the cost of making the proof ineffective in questions of transcendence measures. A quarter of a century later, we know that effectivity is not lost by avoiding derivatives. The earliest work in this direction is [53]; further developments have already been quoted ([31] and [64] for instance). However it is clear that a lot of work is still to be performed in this direction, and plenty of results are waiting to be unraveled by future generations of mathematicians.

Recent references

We give only a few references to papers or books which have been published during the last 10 years: [11, 32, 66, 13, 67, 44, 12] and [68]. They contain further references to related works.

References

[1] Alaoglu, L., Erdős, P. On highly composite and similar numbers, *Trans. Amer. Math. Soc.*, **56** (1944), 448–469.

[2] Baker, A. *Transcendental Number Theory,* Cambridge Univ. Press, 1975; 2nd ed. 1979.

[3] Baker, A. The theory of linear forms in logarithms, Chap.1 of: *Transcendence Theory: Advances and Applications*, Proc. Conf. Cambridge 1976, ed. A.Baker and D.W.Masser, Academic Press (1977), 1–27.

[4] Balasubramanian, R., Ramachandra, K. Transcendental numbers and a lemma in combinatorics, Proc. Sem. Combinatorics and Applications, Indian Stat. Inst., (1982), 57–59.

[5] Bertrand, D., Masser, D.W. Linear forms in elliptic integrals, *Invent. Math.*, **58** (1980), 283–288.

[6] Bombieri, E. Algebraic values of meromorphic maps, *Invent. Math.*, **10** (1970), 267–287; **11** (1970), 163–166.

[7] Bombieri, E., Lang, S. Analytic subgroups of group varieties, *Invent. Math.*, **11** (1970), 1–14.

[8] Brownawell, W.D., Kubota, K.K. The algebraic independence of Weierstrass functions and some related numbers, *Acta Arith.*, **33** (1977), 111–149.

[9] Chudnovsky, G.V. *Contributions to the theory of transcendental numbers*, Mathematical Surveys and Monographs **19**, Amer. Math. Soc., 1984.

[10] Diaz, G. Grands degrés de transcendance pour des familles d'exponentielles en plusieurs variables, *J. Number Theory*, **31** (1989), 1–23.

[11] Diaz, G. La conjecture des quatre exponentielles et les conjectures de D. Bertrand sur la fonction modulaire, *J. Théor. Nombres Bordeaux* **9** (1997), no. 1, p. 229–245.

[12] Diaz, G. Utilisation de la conjugaison complexe dans l'étude de la transcendance de valeurs de la fonction exponentielle, *J. Théor. Nombres Bordeaux* **16** (2004), 535–553.

[13] Fel'dman, N.I., Nesterenko, Yu. V. *Number theory. IV. Transcendental Numbers*, Encyclopaedia of Mathematical Sciences, **44**. Springer-Verlag, Berlin, 1998.

[14] Gel'fond, A.O. On Hilbert's seventh problem, *Dokl. Akad. Nauk SSSR*, **2** (1934), 1–3 (in Russian) and 4–6 (in French), Sur le septième problème de Hilbert, *Izv. Akad. Nauk SSSR*, **7** (1934), 623–630.

[15] Gel'fond, A.O. *Transcendental Number Theory*, Moscow, 1952, English transl. Dover Publ., N.Y., 1960.

[16] Gramain, F., Mignotte, M. Fonctions entières arithmétiques, *Approximations diophantiennes et nombres transcendants*, Luminy 1982, Progress in Math. **31**, Birkhäuser 1983, 99–124.

[17] Gramain, F., Mignotte, M., Waldschmidt, M. Valeurs algébriques de fonctions analytiques, *Acta Arith.*, **47** (1986), 97–121.

[18] Hirata, N. Approximations simultanées sur les groupes algébriques commutatifs, *Compositio Math.*, **86** (1993), 69–96.

[19] Khassa, D.S., Srinivasan, S. A transcendence measure for *e*, *J. Indian Math. Soc.*, **56** (1991), 145–152.

[20] Lang, S. Nombres transcendants, Sém. Bourbaki 18ème année (1965/66), N° 305.

[21] Lang, S. Algebraic values of meromorphic functions, II, *Topology*, **5** (1966), 363–370.

[22] Lang, S. *Introduction to Transcendental Numbers*, Addison-Wesley 1966.

[23] Lang, S. *Elliptic Curves Diophantine Analysis*, Grund. der math. Wiss., **231**, Springer Verlag (1978).

[24] Laurent, M. Sur quelques résultats récents de transcendance, *Journées arithmétiques Luminy* 1989, Astérisque, **198–200** (1991), 209–230.

[25] Laurent, M. Hauteurs de matrices d'interpolation, *Approximations Diophantiennes et Nombres Transcendants*, Luminy 1990, éd. P. Philippon, de Gruyter 1992, 215–238.

[26] Laurent, M. Linear forms in two logarithms and interpolation determinants. *Acta Arith.* **66** (1994), no. 2, 181–199.

[27] Laurent, M., Mignotte, M., Nesterenko, Yu. V. Formes linéaires en deux logarithmes et déterminants d'interpolation. *J. Number Theory* **55** (1995), no. 2, 285–321.

[28] Loxton, J.H. Some problems involving powers of integers, *Acta Arith.*, **46** (1986), 113–123.

[29] Mahler, K. *Lectures on Transcendental Numbers*, Lecture Notes in Math., **546**, Springer-Verlag 1976.

[30] Mignotte, M., Waldschmidt, M. Approximation des valeurs de fonctions transcendantes Koninkl. *Nederl. Akad. van Wet. Proc.*, Ser.A, **78** (=Indag. Math., **37**), (1975), 213–223.

[31] Mignotte, M., Waldschmidt, M. Linear forms in two logarithms and Schneider's method *Math. Ann.*, **231** (1978), 241–267.

[32] Nesterenko, Yu. V., Philippon, P. *Introduction to algebraic independence theory*, Yuri V.Nesterenko and Patrice Philippon Eds, Instructional conference (CIRM Luminy 1997). Lecture Notes in Math., **1752**, Springer, Berlin-New York, (2001).

[33] Philippon, P. Critères pour l'indépendance algébrique, *Publ. Math. I.H.E.S.*, **64** (1986), 5–52.

[34] Philippon, P. Nouveaux aspects de la transcendance, Journées Arithmétiques (Bordeaux, 1993). *J. Théor. Nombres Bordeaux* **7** (1995), no. 1, 191–218. See also: Transcendance sur les anneaux diophantiens, Séminaire de Théorie des Nombres de Caen 1992-93, Univ. Caen 1994.

[35] Philippon, P., Waldschmidt, M. Formes linéaires de logarithmes simultanées sur les groupes algébriques commutatifs, Sém. Th. Nombres Paris 1986-87, Birkhäuser Verlag, Progress in Math. **75** (1989), 313–347.

[36] Ramachandra, K. Contributions to the theory of transcendental numbers (I), *Acta Arith.*, **14** (1968), 65–72, (II), id., 73–88.

[37] Ramachandra, K. *Lectures on transcendental numbers*, The Ramanujan Institute, Univ. of Madras, 1969, 72 p.

[38] Ramachandra, K. A note on Baker's method, *J. Austral. Math. Soc.*, **10** (1969), 197–203.

[39] Ramachandra, K. Application of Baker's theory to two problems considered by Erdős and Selfridge, *J. Indian Math. Soc.*, **37** (1973), 25–34.

[40] Ramachandra, K. An easy transcendence measure for *e*, *J. Indian Math. Soc.*, **51** (1987), 111–116.

[41] Ramachandra, K., Shorey, T.N. On gaps between numbers with a large prime factor, *Acta Arith.*, **24** (1973), 99–111.

[42] Ramachandra, K., Srinivasan, S. A note to a paper by Ramachandra on transcendental numbers, *Hardy-Ramanujan Journal*, **6** (1983), 37–44.

[43] Roy, D. Matrices whose coefficients are linear forms in logarithms, *J. Number Theory*, **41** (1992), 22–47.

[44] Roy, D. An arithmetic criterion for the values of the exponential function, *Acta Arith.*, **97** (2001), 183–194.

[45] Schneider, Th. Transzendenzuntersuchungen periodischer Funktionen, (I), Transendenz von Potenzen, *J. reine angew. Math.*, **14** (1934), 65–69.

[46] Schneider, Th. Transzendenzuntersuchungen periodischer Funktionen, (II), Transzendenzeigenschaften elliptischer Funktionen, *J. reine angew. Math.*, **14** (1934), 70–74.

[47] Schneider, Th. Zur Theorie der Abelschen Funktionen und Integrale, *J. reine angew. Math.*, **183** (1941), 110–128.

[48] Schneider, Th. Ein Satz über ganzwertige Funktionen als Prinzip für Transzendenzbeweise, *Math. Ann.*, **121** (1949), 131–140.

[49] Schneider, Th. *Einführung in die transzendenten Zahlen*, Springer Verlag 1957, trad. franç.,*Introduction aux Nombres Transcendants*, Paris, Gauthier-Villars.

[50] Shorey, T.N. Linear forms in the logarithms of algebraic numbers with small coefficients, *J. Indian Math. Soc.*, **38** (1974), 271–292.

[51] Shorey, T.N. On gaps between numbers with a large prime factor, (II), *Acta Arith.*, **25** (1974), 365–373.

[52] Shorey, T.N. On linear forms in the logarithms of algebraic numbers, *Acta Arith.*, **30** (1976), 27–42.

[53] Srinivasan, S. On algebraic approximations to 2^{π^k} $(k = 1, 2, 3, \ldots)$, *Indian J. Pure Appl. Math.*, **5** (1974), 513–523.

[54] Srinivasan, S. On algebraic approximations to 2^{π^k} $(k = 1, 2, 3, \ldots)$, (II), *J. Indian Math. Soc.*, **43** (1979), 53–60.

[55] Wakabayashi, I. Algebraic values of functions on the unit disk, *Proc. Prospects of Math. Sci.*, World Sci. Pub., (1988), 235–266.

[56] Waldschmidt, M. Indépendance algébrique des valeurs de la fonction exponentielle, *Bull. Soc. Math. France*, **99** (1971), 285–304.

[57] Waldschmidt, M. Propriétés arithmétiques des valeurs de fonctions méromorphes algébriquement indépendantes, *Acta Arithm.*, **23** (1973), 19–88.

[58] Waldschmidt, M. *Nombres transcendants*, Lecture Notes in Math., **402**, Springer–Verlag (1974), 277 p.

[59] Waldschmidt, M. Pólya's Theorem by Schneider's method, *Acta Math. Acad. Sci. Hungar.*, **31** (1978), 21–25.

[60] Waldschmidt, M. Fonctions entières et nombres transcendants , 103è Cong. Nat. Soc. Savantes, Nancy 1978, Sciences, Fasc. V, 303-317.

[61] Waldschmidt, M. Algebraic independence of values of exponential and elliptic functions, *J. Indian Math. Soc.*, **48** (1984), 215–228.

[62] Waldschmidt, M. On the transcendence methods of Gel'fond and Schneider in several variables, Chap. 24 de : New Advances in Transcendence Theory, (ed. A. Baker), Proc. Conf. Durham 1986, (ed. A.Baker), Cambridge Univ. Press (1988), 375–398.

[63] Waldschmidt, M. Algebraic values of analytic functions, International Symposium on Algebra and Number Theory, Silivri (Istanbul), 1989, Doḡa – Tr. J. of Math., **14** (1990), 70–78. Résumé en turc : Analitic fonksiyonların cebirsel değerleri.

[64] Waldschmidt, M. *Linear independence of logarithms of algebraic numbers* , The Institute of Mathematical Sciences, Madras, IMSc Report No **116**, (1992), 168 p.

[65] Waldschmidt, M. Densité de points rationnels sur un groupe algébrique, Experiment. Math. **3** (1994), no. 4, 329–352, Errata, ibid., **4** (1995), no. 3, 255.

[66] Waldschmidt, M. Integer valued functions on products, *J. Ramanujan Math. Soc.*, **12** (1997), no. 1, 1–24 .

[67] Waldschmidt, M. *Diophantine Approximation on Linear Algebraic Groups. Transcendence Properties of the Exponential Function in Several Variables*, Grundlehren der Mathematischen Wissenschaften **326**. Springer-Verlag, Berlin-Heidelberg, 2000.

[68] Waldschmidt, M. Variations on the Six Exponentials Theorem, with an appendix by H. Shiga: Periods of a Kummer surface. International Conference on Algebra and Number Theory Hyderabad 2003, to appear.

[69] Yu Kunrui, Linear forms in elliptic logarithms, *J. Number Theory*, **20** (1985), 1–69.

Michel Waldschmidt
Université P. et M. Curie (Paris VI),
Institut de Mathématiques CNRS UMR 7586,
175, rue du Chevaleret, F–75013 Paris.

e-mail: miw@math.jussieu.fr
http://www.math.jussieu,fr/~miw/

The Riemann Zeta Function and Related Themes – 2006, pp. 181–191

Primes in almost all short intervals and the distribution of the zeros of the Riemann zeta-function

Alessandro Zaccagnini

Dedicated to Professor K. Ramachandra on his 70th birthday

Abstract

We study the relations between the distribution of the zeros of the Riemann zeta-function and the distribution of primes in "almost all" short intervals. It is well known that a relation like $\psi(x) - \psi(x - y) \sim y$ holds for almost all $x \in [N, 2N]$ in a range for y that depends on the width of the available zero-free regions for the Riemann zeta-function, and also on the strength of density bounds for the zeros themselves. We also study implications in the opposite direction: assuming that an asymptotic formula like the above is valid for almost all x in a given range of values for y, we find zero-free regions or density bounds.

1 Introduction

We investigate the relations between the distribution of the zeros of the Riemann zeta-function and the distribution of primes in "almost all" short intervals. Here and in the sequel "almost all" means that the number of integers $n \leq x$ which do not have the required property is $o(x)$ as $x \to \infty$. It is known since Hoheisel's work in the 1930's [8] that a relation of the type

$$\psi(x) - \psi(x - y) \sim y \tag{1}$$

holds in a certain range for y whose width depends on the strength of both zero-free regions and density estimates for the zeros of the zeta function. The same relation is true, obviously in a much wider range for y, if we deal with the same problem but only almost everywhere, that is, allowing a small set of exceptions. Here we show that a partial converse is also true: if (1) holds almost everywhere in a wide region of values for y, then the zeros of the Riemann zeta-function can not be too dense near $\sigma = 1$, nor too close to the same line, in a fairly strong quantitative sense. The corresponding relation between the error term in the Prime Number Theorem and the width of the zero-free region for the zeta-function is a classical result of Turán [19].

2 Prime numbers in all short intervals

For $x \geq 2$ let π and ψ denote the usual Chebyshev functions and set

$$R_1(x) \stackrel{\text{def}}{=} \sup_{u \in [2,x]} \left| \pi(u) - \int_2^u \frac{dt}{\log t} \right|$$

$$R_2(x) \stackrel{\text{def}}{=} \sup_{u \in [2,x]} \left| \psi(u) - u \right|.$$

The Prime Number Theorem (PNT), in the sharpest known form due to Vinogradov and Korobov (see Titchmarsh [18, §6.19]), asserts that

$$R_j(x) \ll x \exp\left\{-c_j(\log x)^{3/5}(\log\log x)^{-1/5}\right\}$$

for suitable $c_j > 0$, $j = 1, 2$. The additivity of the main terms for both π and ψ suggests the truth of the statements

$$\pi(x) - \pi(x-y) \sim \int_{x-y}^{x} \frac{\mathrm{d}t}{\log t} \sim \frac{y}{\log x}, \tag{2}$$

$$\psi(x) - \psi(x-y) \sim y, \tag{3}$$

for $y \le x$ (provided, of course, that y is not too small with respect to x), and these relations are trivial for large y, that is $y/(R_j(x)\log x) \to \infty$. The unproved assumption that all the zeros $\beta + \mathrm{i}\gamma$ of the Riemann zeta-function with real part $\beta \in (0,1)$ actually have $\beta = \frac{1}{2}$ is known as the Riemann Hypothesis (RH). It implies that $R_j(x) \ll x^{1/2}(\log x)^j$ for $j = 1, 2$, and this is essentially best possible since Littlewood [11] proved in 1914 that

$$\limsup_{x\to\infty} \frac{R_1(x)\log x}{x^{1/2}\log\log\log x} > 0.$$

Assuming the RH, Selberg [17] proved in 1943 that (2) holds, provided that $y/(x^{1/2}\log x) \to \infty$. The best result to-date is due to Heath-Brown [6], and is described in the following section.

2.1 Density Estimates

The connection with density estimates arises from the following well-known fact: if there exist constants $C \ge 2$ and $B \ge 0$ such that

$$\begin{aligned} N(\sigma, T) &\stackrel{\text{def}}{=} |\{\varrho = \beta + \mathrm{i}\gamma : \zeta(\varrho) = 0, \beta \ge \sigma, |\gamma| \le T\}| \\ &\ll T^{C(1-\sigma)}(\log T)^B \end{aligned} \tag{4}$$

for $\sigma \in [\frac{1}{2}, 1]$, then it is comparatively easy to prove that, for any fixed $\varepsilon > 0$, both (2) and (3) hold in the range

$$y \ge x^{1-C^{-1}+\varepsilon}. \tag{5}$$

Huxley [9] showed that (4) holds with $C = 12/5$, and therefore $7/12$ is an admissible exponent in (5). Heath-Brown [6] has improved on (5), showing that if $\varepsilon(x) > 0$ for all $x \ge 1$, then a quantitative form of (2) holds for

$$x^{7/12-\varepsilon(x)} \le y \le \frac{x}{(\log x)^4}$$

provided that $\varepsilon(x) \to 0$ as $x \to \infty$. In fact, he proved that

$$\pi(x) - \pi(x-y) = \frac{y}{\log x}\left(1 + O\left(\varepsilon^4(x)\right) + O\left(\left(\frac{\log\log x}{\log x}\right)^4\right)\right).$$

The Density Hypothesis (DH) is the conjecture that (4) holds with $C = 2$ and some $B \ge 0$; in view of the Riemann–von Mangoldt formula (see Titchmarsh [18, Theorem 9.4]), the Density Hypothesis is optimal apart from the value of B, and it yields the exponent $1/2$ in (5). Thus DH and RH have almost the same consequence as far as this problem is concerned.

2.2 Negative results

Maier [12] startled the world in 1985 by proving that (2) is false for $y = (\log x)^\lambda$ for any fixed $\lambda > 1$. Hildebrand & Maier [7] improved on this in 1989 but their result is too complicated to be stated here. Later Friedlander, Granville, Hildebrand & Maier [3] extended these results to arithmetic progressions.

3 Prime numbers in almost all short intervals

For technical convenience we define the Selberg integral by means of

$$J(x,\theta) \stackrel{\text{def}}{=} \int_x^{2x} |\Delta(t,\theta)|^2 \, dt,$$

where $\Delta(x,\theta) := \psi(x) - \psi(x - \theta x) - \theta x$. The natural expectation is that the relation

$$J(x,\theta) = o\big(x^3\theta^2\big) \tag{6}$$

holds in a much wider range for $y = \theta x$ than (5), since we now allow some exceptions to (3). We remark that, when $x^{\varepsilon-1} \le \theta \le 1$, the Brun–Titchmarsh inequality (see for example Montgomery & Vaughan [13]) yields $J(x,\theta) \ll_\varepsilon x^3\theta^2$. Ingham [10] proved in 1937 that (6) holds for $x^{-2C^{-1}+\varepsilon} \le \theta \le 1$. Hence (6) is known to hold in the ranges

$$\begin{cases} x^{-5/6+\varepsilon} \le \theta \le 1 & \text{unconditionally, and} \\ x^{-1+\varepsilon} \le \theta \le 1 & \text{assuming the DH.} \end{cases} \tag{7}$$

Assuming the RH, Selberg [17] gave the stronger bound

$$J(x,\theta) \ll x^2\theta \log^2(2\theta^{-1}),$$

uniformly for $x^{-1} \le \theta \le 1$. Goldston & Montgomery [5], using both the RH and the Pair Correlation Conjecture, showed that if $y = x^\alpha$ then

$$J(x, yx^{-1}) \sim (1-\alpha)xy\log x$$

uniformly for $0 \le \alpha \le 1-\varepsilon$, and Goldston [4], assuming the Generalized Riemann Hypothesis, gave the lower bound

$$J(x, yx^{-1}) \ge \left(\frac{1}{2} - 2\alpha - \varepsilon\right) xy \log x$$

uniformly for $0 \le \alpha < \frac{1}{4}$.

For technical reasons, our result is stated in terms of a modified Selberg integral with the function π in place of ψ, and represents the improvement in the known range of validity for y corresponding to Heath-Brown's result quoted above. The detailed proof can be found in [21].

Theorem 1. *If $x^{-5/6-\varepsilon(x)} \le \theta \le 1$, where $0 \le \varepsilon(x) \le \frac{1}{6}$ and $\varepsilon(x) \to 0$ when $x \to \infty$, then*

$$I(x,\theta) \stackrel{\text{def}}{=} \int_x^{2x} \left| \pi(t) - \pi(t-\theta t) - \frac{\theta t}{\log t} \right|^2 dt \ll \frac{x^3\theta^2}{(\log x)^2} \left(\varepsilon(x) + \frac{\log\log x}{\log x} \right)^2$$

In particular, the assumptions imply that $I(x,\theta) = o\big(x^3\theta^2(\log x)^{-2}\big)$. A simple consequence is that for θ in the above range, the interval $[t-\theta t, t]$ contains the expected number of primes $\sim \theta t(\log t)^{-1}$ for almost all integers $t \in [x, 2x]$.

3.1 Outline of the proof

Since the details of the proof are rather complicated, we start with a weaker, but somewhat easier, result and deal with J instead of I. First we briefly sketch the classical proof of

$$J(x,\theta) = \int_x^{2x} \left|\psi(t) - \psi(t-\theta t) - \theta t\right|^2 \, \mathrm{d}t = o\big(x^3\theta^2\big) \tag{8}$$

uniformly for $x^{-5/6+\varepsilon} \le \theta \le 1$, by means of the Density bound (4) and then give some details of a different proof in the same spirit as Heath-Brown's paper [6]. What follows should be taken with a grain of salt: it is not supposed to be the literal truth, but rather an approximation to it. The interested reader is referred to Saffari & Vaughan [16] for all the details. For brevity, we set $\mathcal{L}_1 := \log x$. By the explicit formula for ψ (see for example Davenport [2, §17]) we have

$$\psi(x) = x - \sum_{|\gamma|<T} \frac{x^\varrho}{\varrho} + O\Big(\frac{x}{T}(\log xT)^2\Big),$$

with the usual convention for the sums over the zeros of the zeta function. Hence

$$J(x,\theta) \ll \int_x^{2x} \Big|\sum_{|\gamma|<T} t^\varrho \frac{1-(1-\theta)^\varrho}{\varrho}\Big|^2 \, \mathrm{d}t + \frac{x^3}{T^2}(\log xT)^4.$$

The last term is harmless provided that $T < x$ and $T^{-1} = o\big(\theta\mathcal{L}_1^{-2}\big)$ which we now assume. We now skip all details until the very last step. The integral is easily rearranged into

$$\sum\sum_{|\gamma_1|<T,\,|\gamma_2|<T} \frac{1-(1-\theta)^{\varrho_1}}{\varrho_1}\,\frac{1-(1-\theta)^{\bar\varrho_2}}{\bar\varrho_2} \int_x^{2x} t^{\varrho_1+\bar\varrho_2} \, \mathrm{d}t.$$

Since $(1-(1-\theta)^\varrho)\varrho^{-1} \ll \theta$, we have that the last expression is

$$\begin{aligned} \ll \theta^2 \sum_{|\gamma|<T} x^{1+2\beta} &= x\theta^2 \sum_{|\gamma|<T} \Big\{2\mathcal{L}_1 \int_{1/2}^{\beta} x^{2\sigma}\,\mathrm{d}\sigma + x\Big\} \\ &= x\theta^2 \Big\{2\mathcal{L}_1 \int_{1/2}^{1} N(\sigma,T)x^{2\sigma}\,\mathrm{d}\sigma + xN\Big(\frac{1}{2},T\Big)\Big\}. \end{aligned}$$

The last summand is negligible if we slightly strengthen our demand to $T = o\big(x\mathcal{L}_1^{-1}\big)$, since $N(\frac{1}{2},T) \ll T\log T$ by the Riemann–von Mangoldt formula. It is clear that we achieve our goal if we can prove that

$$\int_{1/2}^{1} N(\sigma,T)x^{2\sigma}\,\mathrm{d}\sigma = o\big(x^2\mathcal{L}_1^{-1}\big). \tag{9}$$

It is important to stress the fact that simply plugging (4) into this integral does not suffice, since it would only yield the bound $\ll x^2\mathcal{L}_1^B$ for the integral in question. Actually, what is needed for Theorem 1 is a zero-free region for the Riemann zeta-function of the shape

$$\sigma > \sigma_0(x) \stackrel{\text{def}}{=} 1 - \frac{c}{\mathcal{L}_1^A} \tag{10}$$

for some fixed $A \in (0, 1)$, and Korobov & Vinogradov proved that one can take any $A > \frac{2}{3}$ (see Titchmarsh [18, §6.19]). In fact, (10) clearly implies that the upper limit for the integral in (9) can be replaced by $\sigma_0(x)$, and this gives the desired bound provided that T is chosen (optimally) satisfying $T^C = o\left(x^2\mathcal{L}_1^{-4}\right)$, and then it easily follows that (6) holds in the range (7).

3.2 The full result

In order to get the full result given by Theorem 1 we need a more involved argument. We use the Linnik–Heath-Brown identity: For $z > 1$

$$\log(\zeta(s)\Pi(s)) = \sum_{k\geq 1} \frac{(-1)^{k-1}}{k}(\zeta(s)\Pi(s) - 1)^k = \sum_{k\geq 1}\sum_{p\geq z} \frac{1}{kp^{ks}} = \sum_{n\geq 1} \frac{c_n}{n^s},$$

say, where $\Pi(s) = \prod_{p<z}(1 - p^{-s})$. Hence, if $z \leq \frac{1}{2}x$, $\theta \leq \frac{1}{2}$ and $t \in [x, 2x]$ then

$$\pi(t) - \pi(t - \theta t) = \sum_{t-\theta t<n\leq t} c_n + O\left(\theta x^{1/2}\right).$$

For $k \geq 1$ put

$$(\zeta(s)\Pi(s) - 1)^k = \sum_{n\geq 1} \frac{a_k(n)}{n^s},$$

so that $a_k(n) = a^{*k}(n)$, a^{*k} being the k-fold Dirichlet convolution of a_1 with itself, where $a_1(1) = 0$ and $a_1(n) = 0$ unless all prime factors of n are $\geq z$, in which case $a_1(n) = 1$. If $z > x^{1/k_0}$ then $(\zeta(s)\Pi(s) - 1)^k$ has no non-zero coefficient for $n \leq x$ and $k \geq k_0$. We will eventually choose $k_0 = 4$. It is far from easy to approximate $\zeta(s)\Pi(s) - 1$ with Dirichlet polynomials. We have

$$\pi(t) - \pi(t - \theta t) = \sum_{1\leq k<k_0} \frac{(-1)^{k-1}}{k} \sum_{t-\theta t<p\leq t} a_k(n) + O\left(\theta x^{1/2}\right).$$

The goal is to find a function $\mathfrak{M}_k(t)$ which is independent of θ, such that

$$\Sigma_k(t, \theta) \stackrel{\text{def}}{=} \sum_{t-\theta t<p\leq t} a_k(n) = \theta\mathfrak{M}_k(t) + \mathfrak{R}_k(t, \theta),$$

where $\mathfrak{R}_k(t, \theta)$ is "small" in L^2 norm over $[x, 2x]$.

3.3 Reduction to mean-value bounds

We skip the tedious, detailed description of the decomposition into Dirichlet polynomials. Essentially, we can truncate the Dirichlet series for both ζ and Π at height $2x$ without changing the sum we are interested in. Simplifying details, we have

$$\Sigma_k(t,\theta) = \sum_{t-\theta t<p\le t} b_k(n) \qquad \text{where} \qquad \sum_{n\ge 1} \frac{b_k(n)}{n^s} = \prod_{h=1}^{k} f_h(s)$$

for suitable Dirichlet polynomials f_h. Thus we can write $\Sigma_k(t,\theta)$ as a contour integral by means of the truncated Perron formula: neglecting the error term we have

$$\Sigma_k(t,\theta) \sim \frac{1}{2\pi \mathrm{i}} \int_{\frac{1}{2}-\mathrm{i}T_0}^{\frac{1}{2}+\mathrm{i}T_0} (\zeta^*(s)\Pi^*(s)-1)^k \, \frac{t^s-(t-\theta t)^s}{s} \, \mathrm{d}s,$$

the stars meaning that we have truncated the Dirichlet series at $2x$. Here $T_0 = x^{5/6+\beta}$, where $\beta > 0$ is very small but fixed. A short range near 0 of the form $|\Im(s)| \le T_1$ gives the main term of Σ, in the very convenient form $\theta\mathfrak{M}_k(t)$, with $\mathfrak{M}_k(t)$ independent of θ, plus a manageable error term, provided that θ is not too close to 1. For brevity, we do not give the precise definition of T_1 (which depends on the details of the decomposition into Dirichlet polynomials referred to above), and only remark that our final choice satisfies $T_1 = x^{o(1)}$.

We finally have the estimate for the L^2 norm of $\mathfrak{R}_k(t,\theta)$

$$\int_x^{2x} |\mathfrak{R}_k(t,\theta)|^2 \, \mathrm{d}t \ll x^2\theta^2(\log x) \max_{T\in[T_1,T_0]} \int_T^{2T} |\zeta^*(s)\Pi^*(s)-1|^{2k} \, \mathrm{d}t$$

so that we need to prove that the rightmost integral above is $o\big(x(\log x)^{-1}\big)$ uniformly for $T \in [T_1, T_0]$.

The tools needed for completing the proof include the Korobov–Vinogradov zero-free region, the Halász method and Ingham's fourth power moment estimate for the Riemann zeta-function.

Here some of the difficulties arise from the fact that not all Dirichlet polynomial involved have a fixed, bounded number of factors. We had to make a different choice for the Dirichlet polynomials from Heath-Brown, because that choice leads to too large error terms since we have a larger z than Heath-Brown and a much smaller h. This is due to the fact that we need z to be almost $x^{1/3}$, as opposed to $x^{1/6}$. The slight additional difficulty is more than compensated by the fact that we only have to save a little over the estimate given by Montgomery's theorem, since our problem leads naturally to estimating the mean-square of a Dirichlet polynomial. For the details see the author's paper [21].

4 The inverse problem

What can one say about the zeros of the Riemann zeta-function, given bounds for $J(x,\theta)$ (or for $I(x,\theta)$) uniformly in a suitable range for θ? In other words, does a bound for J imply something like a density theorem or a zero-free region for the zeta function? The answer is positive and the general philosophy is that good estimates for J in a sufficiently wide range of uniformity for θ yield good zero-free regions for the zeta function, just as one would expect. It should be remarked, however, that we do not establish a real equivalence between density bounds and bounds for J. This is very clearly illustrated by Corollary 1 below.

Our main result is the following

Theorem 2. *Assume that*

$$J(x,\theta) \ll \frac{x^3\theta^2}{F(\theta x)} \qquad \textit{uniformly for} \qquad G(x)^{-1} \le \theta \le 1 \tag{11}$$

where F and G are positive, strictly increasing functions, unbounded as x tends to infinity. There exist absolute constants $B_0 \ge 1$ and $C_0 \ge 1$ such that if F and G are as above with $G(x) = x^\beta$ for some fixed $\beta \in (0,1]$, then for any $B \ge \max(B_0, \beta^{-1})$ and any $C > C_0$ we have

$$N(\sigma,T) \ll_{B,C} \frac{T^{BC(1-\sigma)}}{\min(F(T^{B-1}),T)}$$

There is a more general form of this Theorem which gives interesting results also in the case $G(x) = o_\varepsilon(x^\varepsilon)$ for every $\varepsilon > 0$: see Corollary 3. We note that the proof gives the numerical bounds $B_0 \le 40000$ and $C_0 \le 2000\log(16e)$, though these values can without doubt be improved upon, and that the Riemann Hypothesis implies that one can take $F(x) = x(\log x)^{-2}$ and $G(x) = x^{1-\varepsilon}$.

Corollary 1. *If* (11) *holds with $F(x) = x^\alpha$ and $G(x) = x^\beta$ for some fixed $\alpha, \beta \in (0,1]$ then*

$$\Theta \overset{\text{def}}{=} \sup\{\operatorname{Re}(\varrho)\colon \zeta(\varrho) = 0\} \le 1 - \frac{1}{C_0}\eta$$

where C_0 is the same constant as in Theorem 2, and

$$\eta = \begin{cases} B_0^{-1} & \text{if } \beta \ge B_0^{-1} \text{ and } \alpha \ge (B_0-1)^{-1} \\ \alpha(1+\alpha)^{-1} & \text{if } \beta \ge B_0^{-1} \text{ and } \alpha \le (B_0-1)^{-1} \\ \beta & \text{if } \beta < B_0^{-1} \text{ and } \alpha \ge \beta(1-\beta)^{-1} \\ \alpha(1+\alpha)^{-1} & \text{if } \beta < B_0^{-1} \text{ and } \alpha \le \beta(1-\beta)^{-1}. \end{cases}$$

Thus, if one could prove that Theorem 2 holds with $B_0 = 2$ and $C_0 = 1$, then from (11) with $\alpha = \beta = 1 - \varepsilon$ one would recover the Riemann Hypothesis, though a simpler, direct argument suffices (see the begining of §2 in [22]). If instead $F(x) \ll_\varepsilon x^\varepsilon$ for every $\varepsilon > 0$, then our result is the following:

Corollary 2. *If the hypotheses of Theorem 2 hold with $F(x) \ll x^\varepsilon$ for every $\varepsilon > 0$ and $G(x) \ge F(x)$, then for every $B > B_0$ and $t > 2$ the Riemann zeta-function has no zeros in the region*

$$\sigma > 1 - \frac{(B-1)\log F(t)}{BC_0 \log t}.$$

Finally, the general version referred to above yields the following result for a special but interesting choice of slowly growing functions F and G.

Corollary 3. *Let B_0 and C_0 denote the constants in Theorem 2. Then if* (11) *holds with $F(x) = \exp(\log x)^\alpha$ and $G(x) = \exp(\log x)^\beta$ for some fixed $\alpha, \beta \in (0,1]$, then the Riemann zeta-function has no zeros in the region*

$$\sigma > 1 - \frac{1+o(1)}{B_0C_0(\log(2+|t|))^{r(\alpha,\beta)}}$$

where $r(\alpha,\beta) := (1-\min(\alpha,\beta))\beta^{-1}$.

We observe that if $F(x) = (\log x)^A$ then from Corollary 2 we recover Littlewood's zero-free region (which is needed in the proof), while arguing as in the proof of Corollary 3 we can show that one recovers the Korobov–Vinogradov zero-free region, provided that one can take $F(x) = G(x) = \exp\{(\log x)^{3/5}(\log\log x)^{-1/5}\}$.

Recall that $\Delta(x, \theta) := \psi(x) - \psi(x - \theta x) - \theta x$. In order to put our results into proper perspective, we recall that Pintz [14], [15] proved that if $\varrho_0 = \beta_0 + i\gamma_0$ is any zero of the Riemann zeta-function, then

$$\limsup_{x\to\infty} \frac{|\Delta(x,1)|}{x^{\beta_0}} \geq \frac{1}{|\varrho_0|}$$

and also the more precise inequality $\Delta(x, 1) = \Omega(x \exp\{-(1 - \varepsilon)\omega(x)\})$ where $\omega(x) := \min\{(1 - \beta)\log x + \log|\gamma| : \varrho = \beta + i\gamma\}$.

4.1 Sketch of the proof

Since every zero $\varrho = \beta + i\gamma$ of the Riemann zeta-function gives rise to a pole of the function

$$F(s) \stackrel{\text{def}}{=} -\frac{\zeta'}{\zeta}(s) - \zeta(s)$$

we can detect zeros by counting the "spikes" of the function F just to the right of the critical strip. It should be remarked that since ζ is of finite order (as a Dirichlet series), it does not cancel the contribution of the pole of $-\zeta'/\zeta$ at $s = \varrho$. Obviously F is related to the function $\Delta(x, \theta)$ defined above, which is the sum of the coefficients in its Dirichlet series in the interval $(x - \theta x, x]$. The function Δ, in its turn, appears in the Selberg integral. Since we are assuming that $|\Delta(t, \theta)|$ is usually small, that is, it is small in the L^2-norm over the interval $[x, 2x]$, this piece of information can be used to show that the zeta function can not have too many zeros.

More precisely, if the zeta function has a zero in the circle $|s - 1 - it| \leq r$, say, then for a suitable integer k, we have that $F^{(k)}(1 + r + it)$ is "large" in a strong quantitative sense. Taking all zeros into account, we find a lower bound for $N(\sigma, T)$. But the assumption on J can be used to give an upper bound for $|F^{(k)}(s)|$, and therefore for $N(\sigma, T)$.

From now on we write c_j for a suitable positive, absolute, effectively computable constant, $\mathcal{L}_2 = \log T$, $w = 1 + it$ with $2 \leq |t| \leq T$, $r = c_7(1 - \sigma)$, $s_0 = w + r$.

The crucial lower estimate is the following bound.

Lemma 1. *There exist absolute constants c_1 and $c_2 > 0$ with the following property: let $\mathcal{L}_2^{-1} \leq r \leq c_1/(16e)$ and $K \geq c_2 r\mathcal{L}_2$. If the Riemann zeta-function has a zero in the circle $|s - w| \leq r$, then there exists an integer k such that $K \leq k \leq 2K$ and*

$$\frac{1}{k!}\left|F^{(k)}(w + r)\right| \geq \frac{2}{c_1}\left(\frac{c_1}{4r}\right)^{k+1}.$$

This can be proved as Lemma A in Bombieri [1, §6], applying a suitable form of Turán Second Main Theorem (see the Corollary of Theorem 8.1 in Turán [20]; the proof of the latter result yields $c_1 = 1/(8e)$) to the function $k!^{-1}\left|F^{(k)}(w + r)\right|$ using the development for the zeta-function

$$\frac{\zeta'(s)}{\zeta(s)} = \sum_{|\varrho - w|\leq 1} \frac{1}{s - \varrho} + O(\mathcal{L}_2),$$

given by Titchmarsh [18, Theorem 9.6], the Cauchy inequalities for the derivatives of holomorphic functions, the Phragmén–Lindelöf principle and the fact that ζ is of finite order in the half plane $\sigma > \frac{1}{2}$.

Lemma 2. *There exist absolute constants $A_0 \geq 1$, $B_0 \geq 1$ and $C_0 \geq 2$ with the following property. Let $\mathcal{L}_2^{-1} \leq r \leq c_1/(16e)$. If the zeta-function has a zero in the circle $|s - w| \leq r$, then for all $x \geq T^{B_0}$ and $C > C_0$ we have*

$$\int_x^{x^{A_0}} \left| \sum_{n \in [x,y]} \frac{\Lambda(n) - 1}{n^w} \right|^2 \frac{\mathrm{d}y}{y} \gg_C (\log x)^3 x^{-Cr}.$$

The proof of this Lemma is close to Bombieri [1], §6, Lemma B, using our Lemma 1 in place of his Lemma A.

Lemma 3. *Uniformly for $x^{\varepsilon-1} \leq \theta \leq \frac{1}{2}$ we have*

$$\int_x^{2x} \left| \sum_{n \in (t-\theta t, t]} \frac{\Lambda(n) - 1}{n} \right|^2 \frac{\mathrm{d}t}{t} \ll_\varepsilon x^{-3} J(x, \theta) + \theta^4.$$

It is a straightforward application of the Brun–Titchmarsh inequality.

Lemma 4. *For $\tau = \exp(\theta)$ we have*

$$\int_{-\theta^{-1}}^{\theta^{-1}} \left| \sum_{n \in (x,y]} \frac{\Lambda(n) - 1}{n} n^{iu} \right|^2 \mathrm{d}u \ll \theta^{-2} \int_x^y \left| \sum_{n \in (u, \tau u]} \frac{\Lambda(n) - 1}{n} \right|^2 \frac{\mathrm{d}u}{u} + \theta.$$

This follows from Gallagher's Lemma (see Bombieri [1, Theorem 9]).

From Lemma 2 we have

$$\int_x^{x^{A_0}} \int_{\gamma - \frac{1}{2}r}^{\gamma + \frac{1}{2}r} \left| \sum_{n \in (x,y]} \frac{\Lambda(n) - 1}{n^{1+iv}} \right|^2 \mathrm{d}v \frac{\mathrm{d}y}{y} \gg r(\log x)^3 x^{-Cr},$$

for any $C > C_0$, and, summing over zeros,

$$N(\sigma, T) r (\log x)^3 x^{-Cr} \ll r \log T \int_x^{x^{A_0}} \int_{-T-r}^{T+r} \left| \sum_{n \in (x,y]} \frac{\Lambda(n) - 1}{n^{1+iu}} \right|^2 \mathrm{d}u \frac{\mathrm{d}y}{y},$$

since each point of the interval $(-T - r, T + r)$ belongs to at most $c_0 r \mathcal{L}_2$ intervals of type $(\gamma - \frac{1}{2}r, \gamma + \frac{1}{2}r)$, by the Density Lemma in Bombieri [1, §6].

Hence

$$\begin{aligned} N(\sigma, T) &\ll \frac{\log T}{(\log x)^3} x^{Cr} \int_x^{x^{A_0}} \int_{-T-r}^{T+r} \left| \sum_{n \in (x,y]} \frac{\Lambda(n) - 1}{n^{1+iu}} \right|^2 \mathrm{d}u \frac{\mathrm{d}y}{y} \\ &\ll \frac{\log T}{(\log x)^3} x^{Cr} \left\{ \theta^{-2} \int_x^{x^{A_0}} \int_x^y \left| \sum_{n \in (u, \tau u]} \frac{\Lambda(n) - 1}{n} \right|^2 \frac{\mathrm{d}u}{u} \frac{\mathrm{d}y}{y} + \theta \log x \right\} \end{aligned}$$

by Lemma 4 with $T = \theta^{-1}$, $\tau = \exp\theta$. The inner integral is essentially

$$\ll \log x \max_{x \le t \le y} \int_t^{2t} \left| \sum_{n \in (u, \tau u]} \frac{\Lambda(n) - 1}{n} \right|^2 \frac{du}{u} \ll \log x \max_{x \le t \le y} J(t, \tau - 1) t^{-3}$$

by Lemma 3. By our hypothesis and our choice of T, we finally have

$$N(\sigma, T) \ll \frac{\log T}{\log x} x^{Cr} \left\{ \frac{1}{F(\theta x)} + \frac{\theta}{\log x} \right\} \ll \frac{\log T}{\log x} x^{Cr} \left\{ \frac{1}{F(xT^{-1})} + T^{-1} \right\}$$

This is our main estimate, subject to the conditions $x \ge \max(T^{B_0}, G^{-1}(T))$ and $\theta = T^{-1}$, and the proof of the various corollaries is fairly straightforward. For the details see the author's paper [22].

References

[1] E. Bombieri. *Le Grand Crible dans la Théorie Analytique des Nombres*. Societé Mathématique de France, Paris, 1974. Astérisque n. 18.

[2] H. Davenport. *Multiplicative Number Theory*. GTM 74. Springer-Verlag, third edition, 2000.

[3] J. Friedlander, A. Granville, A. Hildebrand, and H. Maier. Oscillation theorems for primes in arithmetic progressions and for sifting functions. *Journal of the American Mathematical Society*, 4:25–86, 1991.

[4] D.A. Goldston. A lower bound for the second moment of primes in short intervals. *Expo. Math.*, 13:366–376, 1995.

[5] D.A. Goldston and H. L. Montgomery. Pair correlation of zeros and primes in short intervals. In *Analytic Number Theory and Diophantine Problems*, pages 183–203, Boston, 1987. Birkhäuser.

[6] D.R. Heath-Brown. The number of primes in a short interval. *J. reine angew. Math.*, 389:22–63, 1988.

[7] A. Hildebrand and H. Maier. Irregularities in the distribution of primes in short intervals. *J. reine angew. Math.*, 397:162–193, 1989.

[8] G. Hoheisel. Primzahlprobleme in der analysis. *Sitz. Preuss. Akad. Wiss.*, 33:1–13, 1930.

[9] M.N. Huxley. On the difference between consecutive primes. *Invent. Math.*, 15:164–170, 1972.

[10] A.E. Ingham. On the difference between consecutive primes. *Quart. J. Math. Oxford*, 8:255–266, 1937.

[11] J.E. Littlewood. Sur la distribution des nombres premiers. *C. R. Acad. Sc. Paris*, 158:1869–1872, 1914.

[12] H. Maier. Primes in short intervals. *Michigan Math. J.*, 32:221–225, 1985.

[13] H.L. Montgomery and R.C. Vaughan. On the large sieve. *Mathematika*, 20:119–134, 1973.

[14] J. Pintz. Oscillatory properties of the remainder term of the prime number formula. In *Studies in Pure Mathematics to the memory of P. Turán*, pages 551–560, Boston, 1983. Birkhäuser.

[15] J. Pintz. On the remainder term of the prime number formula and the zeros of the Riemann zeta-function. In *Number Theory*, LNM 1068, pages 186–197, Noordwijkerhout, 1984. Springer-Verlag.

[16] B. Saffari and R.C. Vaughan. On the fractional parts of x/n and related sequences. II. *Ann. Inst. Fourier*, 27:1–30, 1977.

[17] A. Selberg. On the normal density of primes in small intervals, and the difference between consecutive primes. *Arch. Math. Naturvid.*, 47:87–105, 1943.

[18] E.C. Titchmarsh. *The Theory of the Riemann Zeta–Function*. Oxford University Press, Oxford, second edition, 1986.

[19] P. Turán. On the remainder-term of the prime-number formula. II. *Acta Math. Acad. Sci. Hungar.*, 1:155–166, 1950.

[20] P. Turán. *On a New Method of Analysis and its Applications*. J. Wiley & Sons, New York, 1984.

[21] A. Zaccagnini. Primes in almost all short intervals. *Acta Arithmetica*, 84.3:225–244, 1998.

[22] A. Zaccagnini. A conditional density theorem for the zeros of the Riemann zeta-function. *Acta Arithmetica*, 93.3:293–301, 2000.

Alessandro Zaccagnini
Dipartimento di Matematica,
Università degli Studi di Parma,
Parco Area delle Scienze,
53/a, Campus Universitario
43100 Parma, Italy

email: alessandro.zaccagnini@unipr.it

Made in the USA
Coppell, TX
07 August 2024

35733507R00116